GREAT CAREERS

FOR PEOPLE INTERESTED IN...

Science and Technology

Joanna Grigg

Kogan Page

First published in 1998

Design and concept for the series developed in Canada by Trifolium Books Inc. and Weigel Educational Publishers Limited.

Kogan Page Limited
120 Pentonville Road
London N1 9JN

British Library Cataloguing in Publication Data

A CIP record for this book is available from the British Library.

ISBN 0 7494 2599 7

Typeset by Patrick Armstrong, Book Production Services
Printed and bound in Belgium

Acknowledgements

The author would like to thank the many people who contributed in some way to this book, and in particular those who agreed to be interviewed and photographed. It was a pleasure working with you.

Contents

Featured profiles

Careers at a glance

Shelley Murphy

Breakdown Patrol

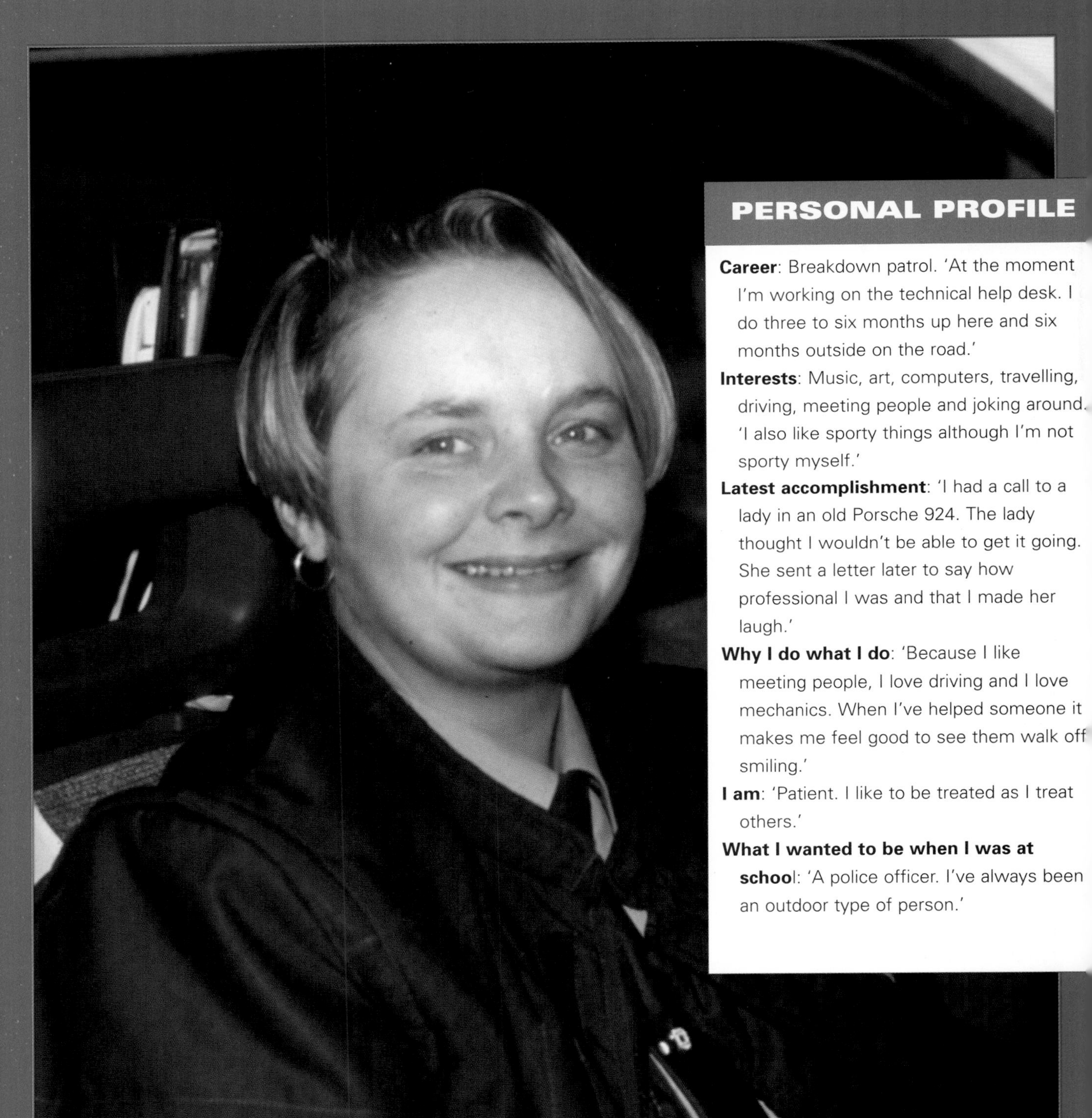

PERSONAL PROFILE

Career: Breakdown patrol. 'At the moment I'm working on the technical help desk. I do three to six months up here and six months outside on the road.'

Interests: Music, art, computers, travelling, driving, meeting people and joking around. 'I also like sporty things although I'm not sporty myself.'

Latest accomplishment: 'I had a call to a lady in an old Porsche 924. The lady thought I wouldn't be able to get it going. She sent a letter later to say how professional I was and that I made her laugh.'

Why I do what I do: 'Because I like meeting people, I love driving and I love mechanics. When I've helped someone it makes me feel good to see them walk off smiling.'

I am: 'Patient. I like to be treated as I treat others.'

What I wanted to be when I was at school: 'A police officer. I've always been an outdoor type of person.'

What a breakdown patrol does

Most of the time cars run well, but there are occasions when motorists have problems – cars break down, or may have flat tyres which the drivers cannot change themselves. They may even run out of petrol. At times like these motorists call a breakdown service such as the Automobile Association (AA). This is one of several national subscription services that offer breakdown insurance; the motorist pays an annual fee and then calls for immediate help when in trouble. The mechanic who comes out to them is known as a patrol. The patrol's aim is to get the motorist safely on the road again but, if this is not possible, he or she will arrange for the car to be towed to a garage for further work.

Shelley's pet hate

'The thing I hate doing is wheel changes; it's so boring. You don't use your brain. The worst are big jeeps. Getting the wheel off the back is fine, it's getting it back on that's the problem. I'm only little and the wheels are bigger than me!'

Shelley Murphy works for the AA as a breakdown patrol. She has a fully equipped van and goes to motorists who need help. She also spends time in the office working on the technical help desk, receiving calls from patrols who are with stranded motorists and need technical help sorting out the problem.

Computer aided

When a member telephones the AA to say there's a problem, the call goes to a central office and, with the help of the computer, operators assign the job to the appropriate patrol. 'Our vans have Mobile Data Terminal systems in them,' explains Shelley. 'I stand by in my van until I get a message on the MDT. It tells you the job, the vehicle, the member's name, location and any phone number.' It also gives an estimated time for Shelley to drive there, do the job and finish. 'Then there'll be another job on the line-up,' she explains. Some jobs take much longer than expected, so Shelley punches in the updated completion time. The line-up is continually updated based on this information.

The Mobile Data Terminal (MDT) system installed in all AA vans helps patrols communicate with headquarters. By punching in certain codes Shelley informs the deployment staff about the type of job she is working on, how long she thinks it will take, and other factors.

When she is helping a member she may have a problem that she cannot solve on her own. ' I could be on a job for ages and not be able to fathom it out,' Shelley explains. 'I call the help desk, and they've got tips and hints on the computer. They scroll through these and say "try this, try that". They're things we might miss or haven't come across before. For instance, the earth wire on the windscreen wiper motor can stop the car on a certain model – you don't always think of things like that.'

Security

Sometimes motorists lock their keys in the car. They call the AA and the patrol needs to be able to break into the car to retrieve them. But patrol vans do not keep information about how to break into different types of cars – in case this information is stolen. The patrol has to telephone the help desk. Only when the help desk is given the patrol's call sign and job number will this security information be given out.

Each van has a work space for dealing with intricate parts and keeping out of the rain. Here, Shelley takes an old distributor to pieces. She uses old parts like these to slot into a broken-down car to ensure that she has diagnosed the problem correctly. Once she is sure, she will buy the correct part on behalf of the member.

The AA has approximately 4600 patrols in the UK. Thirteen of them are women. The most common fault Shelley finds on broken-down vehicles is flat batteries.

All in a day's work

Shelley works a shift system and her day can start early. 'I leave home and drive to the sign-on point, sign on through the MDT and wait for my first job,' she says. Once this has come through and she has accepted it, she drives to the location and pushes the 'arrival' button on her MDT. 'Then I greet the member, ask what the problem is and get a rough idea of the run-up prior to the breakdown. I'm basically asking them what happened.' Then Shelley goes through a number of checks. 'I ascertain whether the vehicle has lost a spark. If there is a spark I go to the fuel side. I check to see if fuel is coming up. You can turn on the ignition and hear the fuel pump prime but it may not be getting up to the plugs.' Not all vehicles need these checks: 'It may be a simple wheel change or flat battery, or the alternator may have gone and the car has been driven on the battery, which has cut out because it's not being recharged.'

Shelley may spend six months on the technical help desk, answering queries from patrols across the country. 'When I first came on the help desk I'd never even seen a wiring diagram before. There is a lot of technical stuff to take in on this job. I thought then I may as well go home. But I get boosted by the other guys at work and it's fine now.'

There are always four or five people working on the technical help desk. They have the use of technical manuals as well as all the data on the computer. They also have each other's experience to call on. Some breakdown problems can involve six minds before they are solved!

Roadside repairs

Some repairs are possible to do by the side of the road but others need more time or equipment. Even the simplest of fan-belt replacements on some vehicles are not feasible by the road. Shelley looks the job up in a timebook, which indicates the estimated time to complete jobs. The trouble is, these times are based on working in a garage. 'If the book says an hour you can add 45 minutes to that. We can't spend that long, we're a breakdown service, we're there to get them going. It's possible to get stuck on a job for two-and-a-half hours sometimes because once you've started it you've got to finish. But if I know it's going to take that long I say: "I can get you going and follow you to the nearest garage, to make sure you don't break down again". Or I tow them there.' If the car cannot be towed, or the member needs to travel further with the car before it is repaired, she will call a recovery van.

'We don't do brakes at the side of the road,' adds Shelley. 'If we changed the brake shoes and there's a leaky calliper, the brakes can fail and it could be very dangerous to drive the car.'

Lateral thinking

As well as diagnosing the problem in the vehicle Shelley has to think about other issues. 'I consider the time factor, whether I'm able to get parts, and whether the member

Shelley in her high-visibility coat. For safety reasons, patrols always wear these when they are working on a job.

has got funds – and is prepared to spend them on parts. We carry some spares but we can't carry everything for every car or we'd be driving juggernauts,' she laughs. She takes all these factors into account before she decides whether to do a full repair, or to get the car going and follow it to a garage, or to tow or recover it. Whichever decision she makes, she endeavours to keep the member happy. 'Sometimes you can't fix a car – maybe the engine's had it – but if the member walks away happy they feel they've got their service. It's all down to customer relations.'

At the end of a job she cleans up: 'I have a black face, black hands, and I may be soaking wet. I put my tools away and get into the van. I write the job up in my personal notebook, punch a seven-digit code into the MDT to see if there's another job in the line-up, then I'm off again.'

'It's nice in the summer and horrible in the winter,' confides Shelley. 'Everyone hates cold weather. We're there in wind, rain or shine. It's nice to be out on the road in the summer. The one thing I detest is the wet, I hate the rain. If you've got the bonnet up and the engine gets soaking wet it won't start anyway. But I'm my own boss,' she adds. 'I get to meet loads of people. I'm in my own environment – I've even got my own little window box,' she laughs, pointing at a plastic rosebud in her van.

Not a good idea...

'People sometimes put diesel in a petrol engine, drive away and then the engine cuts out. I can smell the diesel, and there's nothing I can do. I tow them to a garage, the tank is drained out and it costs them £70 plus a new tankful of petrol.'

Activity

Exploring mechanics

You will need a gadget, or part of one, to take apart and any documentation that goes with it, eg a car manual. (This could be a toy car, a car wiper motor from a scrap yard or an old clock.)

(Note: only dismantle your own gadgets, or get permission if they belong to someone else. Do not dismantle mains electrical items. Use tools only with adult supervision.)

Procedure

Ask yourself these questions about the gadget.

Is this current technology or has it been superseded? If you don't know, how can you find out?

Does the gadget work mechanically, electronically, or in some other way (such as in the case of a CD player, which contains a laser)? Is it a combination of these?

How easily can you take it apart? If you were trying to mend it, what might you think about before you dismantled it? Think of ways to document where each part goes. How easy would it be to put back together?

Now dismantle the item as carefully as possible. Think about: the following.

How you will fit it back together.

If it is broken, can you see why it no longer works? Can you mend it? Is it throwaway technology or designed to be mended?

How does this item fit into the larger picture: is it part of a bigger mechanical item (eg a wiper motor)?

Look at the design. Can you think how it could be improved? Have later models incorporated these design changes?

You may want to draw a diagram of the way it works and how the parts fit together.

Now try to reassemble it. How easy is it? Have you done a good job? How would you do better next time? Is this the sort of work you might enjoy doing?

How to become a breakdown patrol

Shelley didn't want to be a mechanic when she first left school, and she had several jobs, as a beltmaker, cashier and selling car parts. But she always enjoyed using her hands and thinking problems through: 'Things would go wrong at home and I'd have a go at fixing them. I would work on my mother's car – it would take me four times as long as the garage, but I did it. I always liked tinkering with cars and mending things.'

Her previous jobs involved working with people, which she enjoys, but she always felt she wanted more out of life. Eight years ago she got a job in a garage and had an informal apprenticeship training there. She was ready for the entrance tests when she applied to the AA. 'There are three tests,' she explains, 'driving, a practical test and a written test. There's a very high standard.' Part of the reason she was selected was for her customer skills. 'On the initial course they put you in a make-believe customer situation and see how you react,' relates Shelley. 'We are the front line. The only person the members have physical contact with is the patrol.'

After joining the AA she completed a two-week residential training course, and still goes on courses now. 'I came from a garage, servicing and changing brakes, not asking how or why it was diagnosed,' she says, and at first her diagnosis skills were poor. She can ask to be sent on a course or her boss might send her, so she keeps up with the mechanics of new vehicles. 'We go on roadshows to see new vehicles, or to car manufacturers to see the new cars and talk to the technicians there.'

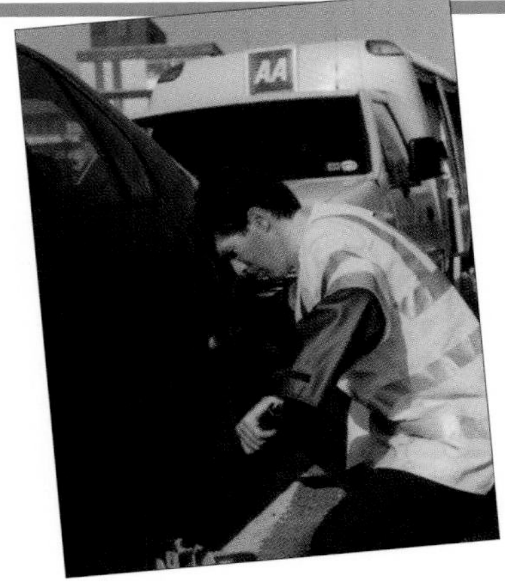

Changing the wheel on a car is a relatively simple job, but it can be troublesome. Some garages fasten the wheel nuts so tightly that special equipment is needed to loosen them. Many motorists simply do not have the strength to fit a spare wheel.

The AA gives priority service to women alone or with children.

Shelley punches details into the MDT at the end of each job, and also keeps a personal log for her own reference.

Is this the career for you?

'You need a good memory,' explains Shelley, 'because what you repair today might crop up again on another vehicle. You also need your wits about you. You must be good technically: at the help desk you get inundated with perhaps 50 or 60 calls a day, with a lot of technical questions. You learn a lot.'

Good communication skills are important and 'you must be able to cope with stresses and tribulations. A member can be in a terrible way and you need to be able to cope with situations like that.' You also need to be able to handle a heavy workload and know your local geography. You must be able to think for yourself and work solely on your own and use your initiative.

'It's not dull,' laughs Shelley. 'No two days are the same. But the main thing,' she adds, 'is to be able to be nice and talk to people. It's all down to how you treat someone, and being able to think that a member you've helped has gone away feeling happy.

Career planning

Making Career Connections

Visit a local garage and ask if you can watch the mechanics at work. Ask them to explain what they are doing, and why.

Find out about local rally teams, drag car racing or other motor sports clubs and ask if you can help out in some way.

See if there is a centre for one of the breakdown services near you. Ask if you can visit and see the control room at work. Ask for details of their training programmes.

Ask your careers adviser about careers in car mechanics. Find out about other types of mechanics jobs too. Send for course details from colleges offering training.

Getting started

Interested in being a breakdown patrol? Here's what you can do now:

1. Keep up all your studies at school. Concentrate on maths, physics, computer science and English as well as practical subjects.
2. Read as many vehicle mechanics books as you can. Gain a good overall picture of the basics of how vehicles work. Look at books on basic electrics and at car magazines too.
3. Attend evening classes in vehicle engineering.
4. Make your own project file on the workings of a car. Learn enough to be able to draw and write the basic principles which are fundamental to all cars. Have additional pages for the extras and more recent additions.
5. Visit car showrooms and shows and ask for brochures of the most recent cars, so you can keep up with the technology.

Related careers

Garage mechanic
Diagnoses problems with cars brought in by customers. Repairs faults and services engines. May specialize in one type or make of vehicle, or one part of the car.

Training officer
Teaches car mechanics to people in college and in work. Keeps up-to-date with the latest technology and passes this on to the students.

Parts dealer
Trades in parts for vehicles. Needs a thorough knowledge of mechanics to understand which part is required and how it is fitted. May drive a van to deliver the parts to garages or patrols.

Car salesperson
Deals in new or used cars. Understands cars and values them for trade-ins. Needs to keep interested in new developments.

Future watch

There are more cars on our roads every year, and this trend looks likely to continue. As long as there are cars, there will always be a demand for the people who look after them. With fewer old cars on the roads, because insurance companies won't pay to mend older models after accidents, patrols and mechanics need to be absolutely up-to-date with all the new developments and keen to see them in action.

Tim York

Research Fellow

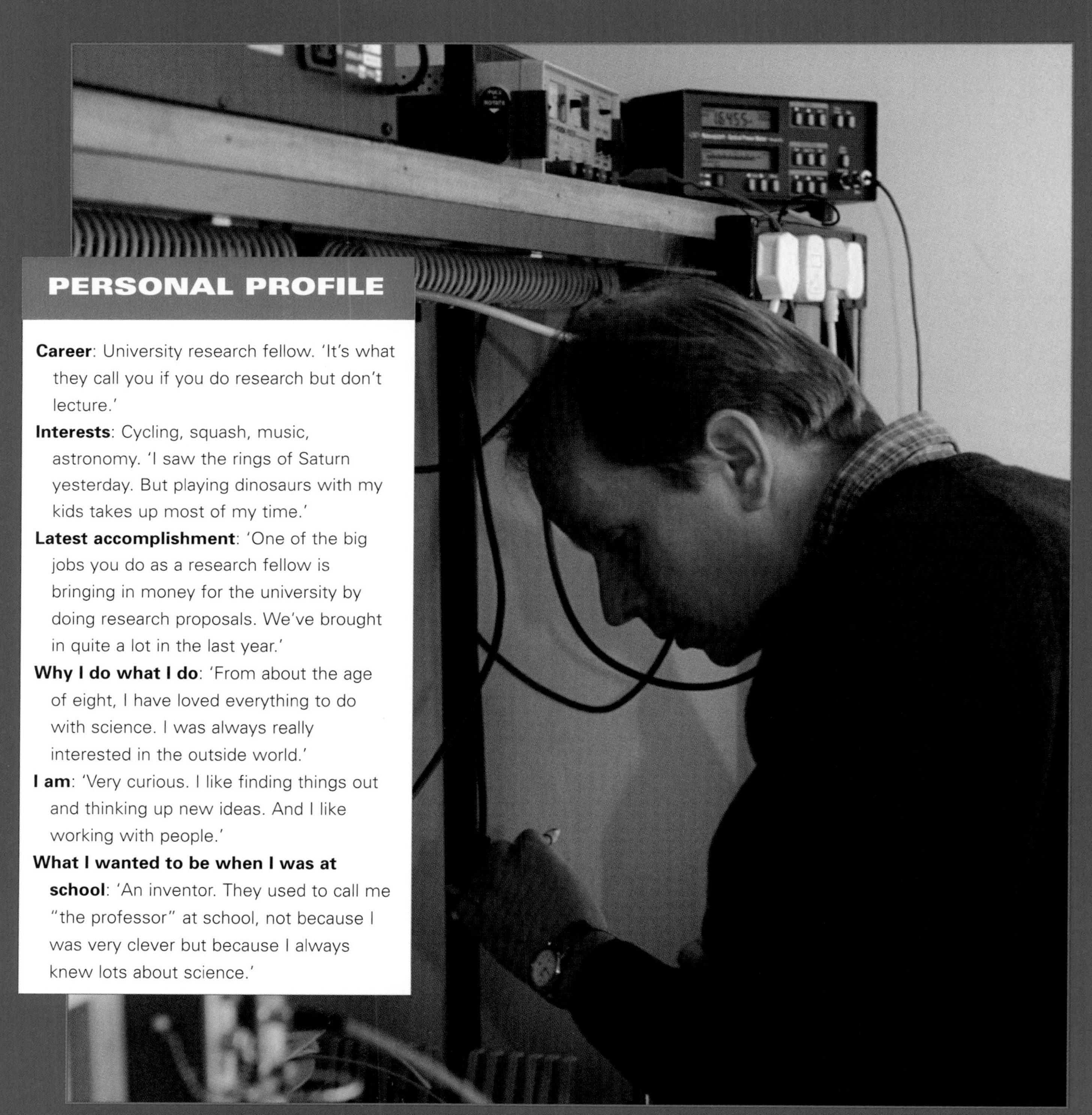

PERSONAL PROFILE

Career: University research fellow. 'It's what they call you if you do research but don't lecture.'

Interests: Cycling, squash, music, astronomy. 'I saw the rings of Saturn yesterday. But playing dinosaurs with my kids takes up most of my time.'

Latest accomplishment: 'One of the big jobs you do as a research fellow is bringing in money for the university by doing research proposals. We've brought in quite a lot in the last year.'

Why I do what I do: 'From about the age of eight, I have loved everything to do with science. I was always really interested in the outside world.'

I am: 'Very curious. I like finding things out and thinking up new ideas. And I like working with people.'

What I wanted to be when I was at school: 'An inventor. They used to call me "the professor" at school, not because I was very clever but because I always knew lots about science.'

What a research fellow does

Most research in UK universities is done by research fellows and PhD students. Lecturers, who are senior, tend to have too many other roles – they will normally discuss the research and input ideas, rather than physically carry it out. Dr Tim York completed his PhD and now works as a research fellow in the department of electronic and electrical engineering at University College London. Although many of his colleagues are engineers he calls himself a physicist. His special field of interest is light.

'There are two types of research in my area,' explains Tim. 'One is pure research, sometimes called "blue sky research". This might be when something interesting has happened in the lab, an interesting observation, for instance, that's difficult to explain or has no obvious technological application. Or it might be a big problem in science that people want to explain, such as the fundamental nature of matter. People want to know, but there's no obvious gain to mankind.

'The other type,' continues Tim, 'is applied research. An example is a problem with liquid crystal displays. To make them work well you've got to put a lot of light through them. Only about four per cent of the light at the back comes through. This means that portable computers, for instance, use far more power than they need and the battery goes flat after a couple of hours.' Tim is currently looking at ways of improving the efficiency of these displays.

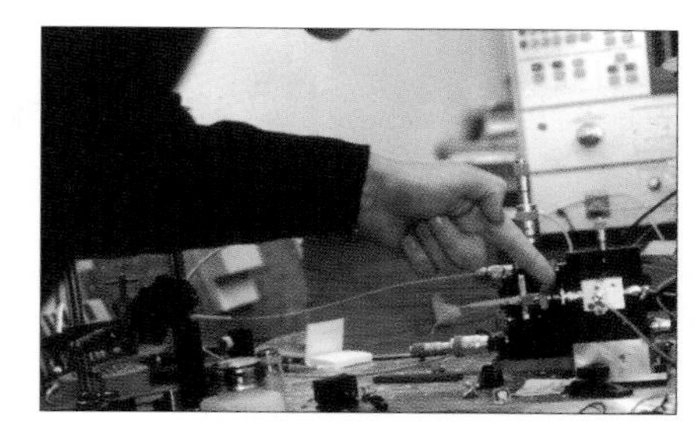

Tim is pointing to the laser component of the VCSEL equipment (the Vertical Cavity Surface Emitting Laser).

Tim at his computer, researching on the Internet. Confidence to use computer technology is an essential requirement of his job.

Who pays?

Tim works solely in applied research. 'Even applied problems are interesting because it's nice to see a solution working,' he says. His research is funded by grants given jointly by companies in industry who will benefit directly from the results, and from research councils. 'It's the way a lot of research is going at the moment,' explains Tim, 'towards a particular problem. It is more difficult to get grants for pure research, which is a shame because our intellectual wealth and long term future depend on blue sky research.'

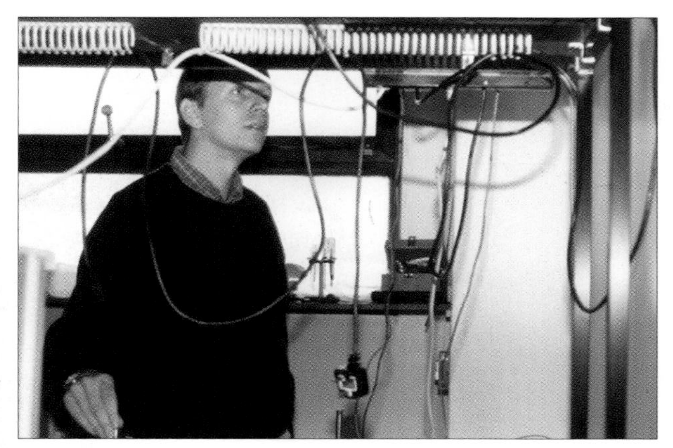

Tim is working in a laboratory specially set up to analyse the bandwidth of laser fibre systems. The research is being undertaken on behalf of a computer manufacturing company.

It's a fact

Modern semiconductor electronics can only be understood in terms of quantum theory developed in the 1930s. In those days, nobody imagined that there would be an application for this esoteric branch of physics.

All in a day's work

Tim cycles into central London each morning and starts his working day by checking his e-mail. He then goes to his laboratory. 'I'll either talk with colleagues about the direction of our work, and plan for the following week,' explains Tim, 'or I'll do some measurements and tests and play with ideas.' Before he can start work in the laboratory he needs to turn on the equipment – large lasers can take half an hour to warm up, check, and tune if they are out of alignment.

'I'm working on two projects at the moment,' explains Tim. 'One involves liquid crystal displays (LCDs).' Here, Tim is placing tiny holograms on the surface of the piece of glass which forms part of an LCD computer monitor. These holograms, he hopes, will redirect the light coming from inside the screen more efficiently than the current system. This also involves placing hundreds of micro lenses on the outside of the liquid crystal. 'I might go into a clean room and make micro lenses,' he explains. 'I coat the surface of a piece of glass with a substance called photo-resist, expose it to a certain pattern of light then put it in the oven. I use the resulting tiny lenses to test my new monitor system.'

It's a fact

Until very recently there were no blue LEDs – light emitting displays – only red, green and yellow. LEDs use less power and need less maintenance than light bulbs, so they are used to make products such as traffic lights. This research breakthrough for a blue LED has meant a far greater range of uses for LEDs.

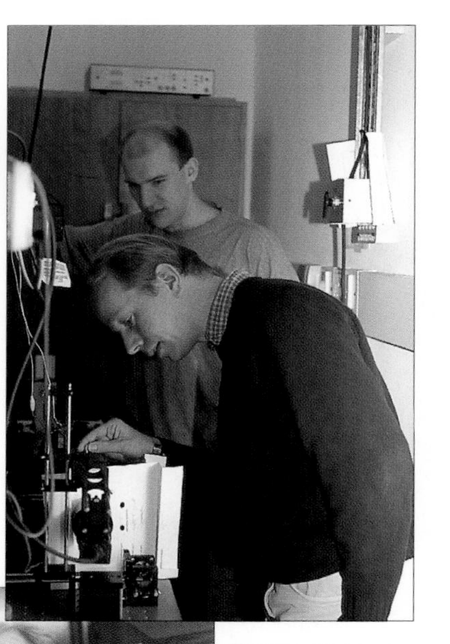

Tim's colleague Lawrence researches into liquid crystals with micro lenses. In Lawrence's laboratory, the two scientists look at the analysis of the phase of distribution of light coming out of a micro lens.

It's a fact

The word 'laser' is an acronym and stands for Light Amplification by Stimulated Emission of Radiation. It originated from one of Einstein's ideas, later developed with quantum mechanics and optical technology. When the first lasers were developed in the 1960s no one knew what to do with them, but now many homes have one – in the CD player.

High speed links

A computer manufacturer wants to build optical fibre links between its computers and monitors, as these allow faster transmission of greater amounts of information than with conventional wires. Future computer monitors may be enormous and flat, and hung on walls in places like airports, with the actual computer a long way from the monitor. It's now getting to the stage where it's difficult to transmit the required amount of information along wires, so Tim is helping design a system to build a cost-effective linkage. This is his second project.

'My projects might involve 90 per cent lab work,' says Tim, 'or only ten per cent.' The rest of the time could be spent on computer modelling or even good old-fashioned mathematical equations. 'I like experimenting and computer modelling – I like playing with things,' he continues. 'Most people who really enjoy science like playing with ideas and physical things.'

Team work

'I couldn't do a lot of these things in an isolated environment,' Tim explains. 'You have got to talk to people, get them to show you things, lend you equipment, chat people up and co-operate. It's an important part of being a scientist. You don't have time to learn everything from scratch.'

Tim spends a lot of time in meetings with his colleagues and his partners in industry. These might involve telling others about his results, planning the next stage of the project, discussing funding and many other issues. He also attends international conferences and looks after the grants. 'I spend a lot of time checking where grant money has got lost in the finance jungle,' he explains. 'I also talk to potential suppliers of industrial equipment and order what I need.' He publishes scientific papers detailing his research results, though not as many as some scientists: 'With my research I often have non-disclosure agreements with companies. Industrial organizations try to hide their discoveries while universities go out of their way to tell everyone about theirs!'

Tim outside University College London. He is lucky to be working in science– particularly in an area in which he has always had a passionate interest.

Five Star Research

Everybody gets marked – even researchers. There is a grading system applied to university departments, looking at their quality of research and teaching. University College London's electronic and electrical engineering has the highest possible five stars, and is consistently within the top five departments of its type in the country.

Activity

Spotting light sources

(Important note: never look directly or indirectly at the sun, even through darkened film. Use a wide bowl of water and look at the sun's reflection instead.)

How many different types of light sources can you find? You need to look around the natural environment as well as your home and school or college.

Try to find an example of each of these:

1. Light emitting diodes
2. Incandescent light
3. Fluorescent light
4. Light created when electrons hit a phosphor
5. Lasers
6. Sodium bulbs
7. Nuclear power
8. Electroluminescence

(answers on page 48)

How to become a research fellow

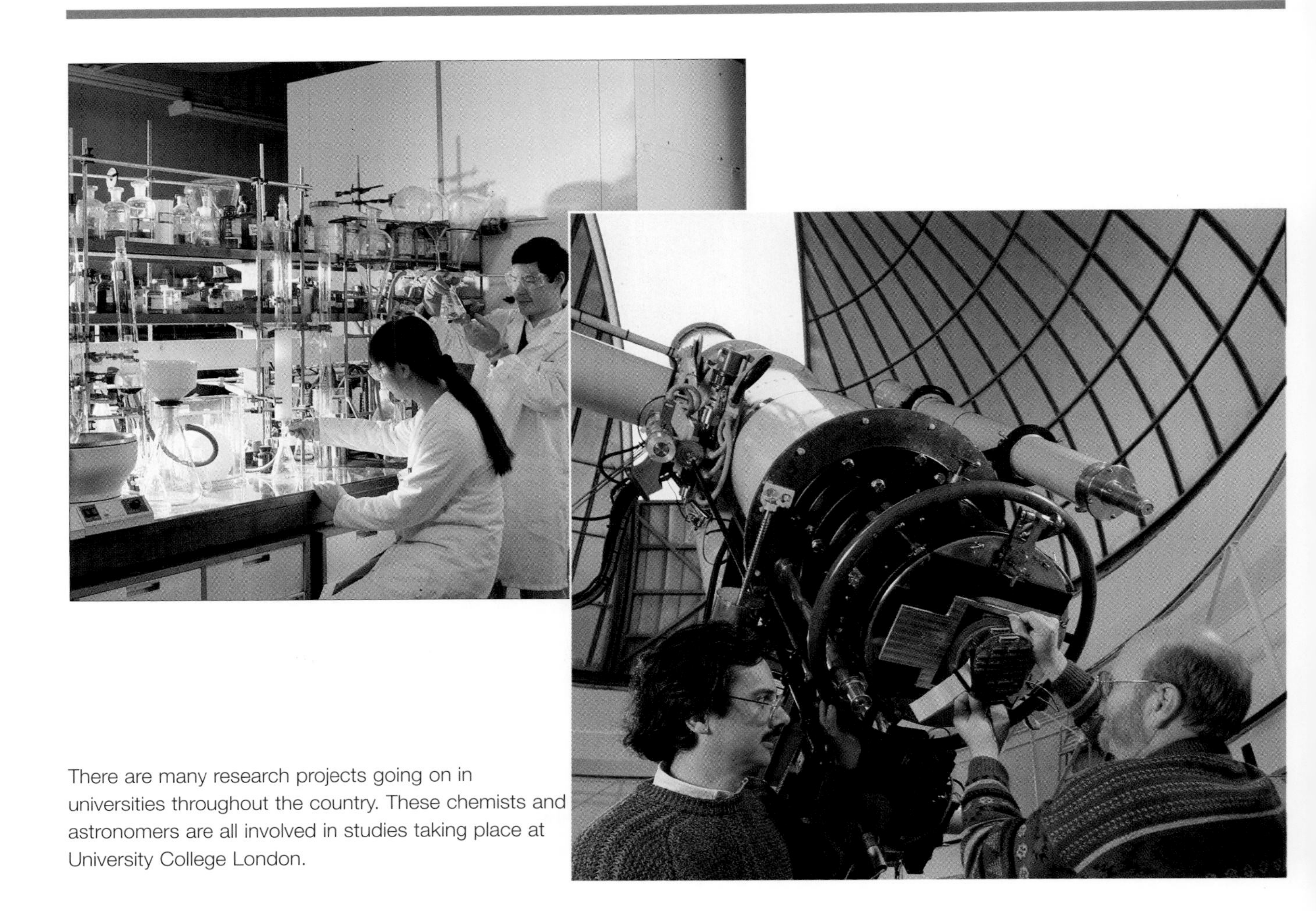

There are many research projects going on in universities throughout the country. These chemists and astronomers are all involved in studies taking place at University College London.

Is this the career for you?

Tim loves his work. 'The main thing I'm interested in is light,' he says. 'Everything is beautiful because of light. Things that manipulate light, such as prisms, are beautiful. They're perfect, precise, and last a long time.' This enthusiasm for his subject is important. 'You've got to be quite academic,' says Tim, 'as well as very interested in the work that you do. You also need to enjoy working both independently and with small groups of people.'

'There's more freedom in your work here than if you work in a company,' explains Tim. 'You can invent your own job if you can interest other people and attract the research funding. You do need to be quite creative. People don't think of science as creative but it is, in the same way that a designer has an idea and finds the best way to put it over. It's also one of the few jobs where you can have a new idea and really see it through. It's very unusual to have a truly new idea though. More often you think you have a truly new idea only to find that someone else has already thought of it.'

Tim appreciates his job.'People are very lucky if they have one overriding interest in their lives, especially if that interest is science: there's work in science.'

Career planning

Making Career Connections

Contact a nearby university which does research in a topic that interests you. Discover this from the prospectus: looking at course content will give you a rough guide to the department's specialisms. Ask if you can visit.

Ask your careers adviser about the best route into an academic career. Find out which universities have the best reputations for work in your field of interest.

Invite a researcher into your college or school to talk about work in research.

Ask local industrial organizations whether they have their own research department. If so, ask whether you can visit or invite a member of staff to talk to your group.

Getting started

Interested in being a research fellow? Here's what you can do now.

1. Concentrate on all your studies, not just sciences. Researchers need good writing and listening skills among many others.
2. Investigate computer modelling. Learn as much as you can about state-of-the-art computers. Make sure your keyboard skills are excellent.
3. Carry out your own science investigations. Learn about method and accurate recording of results as well as the idea and results themselves.
4. Get involved in school, college or local science organizations. Offer to present results of any research you are doing at the moment.

Related careers

Here are some related careers you may want to look into.

Research technician
Assists research projects in university and commercial laboratories.

Science editor
Works on one of the scientific journals. Needs a science background to understand the papers and prepare them for publication.

Science journalist
Works for a science magazine or for a newspaper. Monitors research findings and selects the most topical and interesting for publication in everyday language.

Lecturer
Teaches undergraduates and supervises research students in a university. Produces own research. Brings in research funding.

Future watch

There will always be a need for research, although it is often among the first things to be cut in a recession. The current emphasis on applied research will continue, with funding increasingly available through industrial organizations, which have disbanded their own research departments and instead 'contract' work to departments in universities. Only the richest nations can afford research, so the location of well-funded research will move as different nations take the economic lead.

Christine Caldwell

Women's Health Worker

PERSONAL PROFILE

Career: Women's health worker. 'Women need to see their family doctor more than men. I am a bit like a nurse practitioner but I only see women patients. I help take the workload off the doctors'

Interests: Cycling, cinema, dancing, going out with friends, reading and cooking. 'I like riding my bike because I like the exercise, and cycling with other people for the social side.'

Latest accomplishment: Getting this job. 'When I saw the job advertised I thought I had a very small chance of getting it.'

Why I do what I do: 'I enjoy being part of a team where we all share equal responsibility for the running of the place, and I enjoy helping people sort out problems that are related to how their bodies work.'

I am: 'The sort of person who cares about relationships with people, and I am anti-authoritarian. I think a lot about things.'

What I wanted to be when I was at school: A social worker.

What a women's health worker does

Everyone in this country should be registered with a family doctor, or general practitioner (GP). A GP's surgery, or practice, needs other staff as well as the doctors themselves and these may include receptionists, nurses, physiotherapists and administrators. Christine Caldwell works in one of these practices as a women's health worker. 'We believe in helping people to take responsibility for their own health,' says Christine. 'We encourage them to look at ways they can make changes in their lives that will improve their health.'

When patients arrive at the surgery they are handed their notes. Christine needs to find these. The staff at Christine's practice believe that patients should have automatic access to their own medical records.

Many practices are changing their attitudes and staffing as we become more aware of the need for preventive medicine. One major change is the relatively new role of nurse practitioner, a qualified nurse who is autonomous and makes certain medical diagnoses and treatment decisions. Christine's job is like this, although she only works with women patients and treats specifically 'women's' problems.

Christine takes a patient's blood pressure.

Working as a collective

'I enjoy working with women,' says Christine, 'And I'm very interested in the way that women's bodies work. In traditional practices, if a woman is having problems with family planning the nurse would refer her to the doctor. In practices with a nurse practitioner or women's health worker, we can discuss the problem with the patient and make decisions together.' Christine is able to give patients longer appointment times than the doctors: 'People coming in for one thing often want to talk about something else too,' explains Christine.

Although many of her patients may not realize it, Christine's practice is not like most others; it is run as a collective. 'A collective is a group of people who jointly own and run whatever business it is,' explains Christine. 'Everyone has equal responsibility for the day-to-day running of the service, so power is shared equally. Everybody is regarded as of equal value, and that applies to how much we're paid.' All the people working with Christine share the administrative work such as staffing the reception desk. 'And most people treat the receptionist in a very different way than they would treat the doctors,' laughs Christine. 'We're very much like an average surgery from the patient's point of view,' she adds. 'It's not perfect or ideal or without its problems or stresses but most of us would say that it works.'

All in a day's work

Christine's days vary according to the surgery she is running and whether she has administrative duties. Her hours vary too, and all staff are able to choose their own hours of working to fit in with their needs and the needs of the patients and the practice. 'When I get in I have a cup of coffee,' says Christine, 'and check my in-tray. I'm responsible for some of the administration so I might write to a patient or write a referral letter to the hospital.' From mid-morning she takes over the reception desk. 'This means greeting patients, giving them their notes and answering the phone. It can be stressful, especially when patients are very demanding – most of us dislike working on reception. It's busy, but I cope!"

Christine calls up information on her computer.

Women's health problems

After a break for lunch Christine finds the notes for patients she will see in the afternoon. 'I like to look through the notes,' explains Christine, 'I like to remind myself what I talked about with patients on their previous visit.' She tidies the clinic and checks that all her equipment is ready. Then she sees patients until early evening: 'It depends on why people have come – sometimes it takes a long time to finish the clinic.' She might see patients about contraception, infertility problems, pre-conceptual care, period problems such as pre-menstrual tension, cervical smears, the menopause and HRT.

Christine needs a sound understanding of the science behind the medical problems she encounters. She needs to make important decisions, and has been trained specifically in each area she covers, in addition to her own scientific background and experience.

Cristina, one of the practice doctors, chats to a patient.The practice is run as a collective and everyone working there has an equal responsibility for its day-to-day management.

Keeping up-to-date

'A lot of the things I do are routine, such as family planning and smears. These can take a long time, so this clears the doctors' time for other patients.'

Afterwards she clears up, sterilizes the equipment, tidies up, makes any necessary phone calls, then leaves. 'Sometimes I'm tired,

Christine removes sterilized equipment from the autoclave before the surgery opens.

depending on what patients have come for and whether there have been any major problems. Sometimes I don't feel I can answer their questions very well and feel I need to do some reading or talk to one of the doctors.' Christine spends time reading articles in the medical press and keeping up-to-date on her specialist subjects.

It's a fact

More and more GP practices are employing nurse practitioners to take the workload off the doctors, although hardly any have women's health workers like Christine.

A large number of women aren't aware of the potential advantages of hormone replacement therapy (HRT) to help treat menopausal symptoms, although many women doctors use it themselves. The doctors often don't have the time to explain complex issues such as HRT, which is the subject of controversial debate. Many patients stop taking it within the first three months because of side effects that haven't been properly explained to them.

It's a fact

Preventive medicine is much more on the medical agenda now. Awareness of diseases such as testicular cancer mean that is detected earlier in young men and is more effectively treated.

Men often have a different approach to their health. They can be more reluctant than women to seek medical advice. Research has shown that this may be one of the reasons for men's shorter life expectancy.

Activity

Researching research

Anyone working in science or technology has to keep up-to-date with the latest developments in their field. This research is an important – and very interesting – part of the work

You will need:

To decide upon an area of science or technology that interests you.

Access to a library, the Internet, research journals or an expert in that field

Procedure

Think about your knowledge of the area you have chosen. How broad is your knowledge? How old are the facts you have been taught? How does your limited knowledge fit into the overall knowledge base of that topic – do you think you know ten per cent of current findings? Fifty per cent perhaps?

1. Make notes about what you already know.
2. Find out how old that knowledge is likely to be – ask your teacher, look at the publication date in your textbook, ask people who are also interested.
3. Think about what new scientific research might have been undertaken since. Isolate keywords to use in your own research. If you are interested in women's health, for instance, keywords might include contraception, cervix or HRT.
4. Look up relevant topics wherever you can – on the Internet or in science journals, for instance – using your keywords. Ask people working in science and technology.
5. Have you increased your knowledge? Can you think how you might be able to apply this knowledge in a scientific or technical job? Have you increased your research skills? Which is more important – having research skills or knowing all the latest discoveries?

How to become a women's health worker

After studying human sciences at university, Christine became an information officer with a poisons information service. 'I spent seven years there,' recalls Christine, 'then I got fed up with just talking to people on the phone all the time – I thought I'd like to see some real patients!' She took a graduate nursing course. 'I trained as a nurse partly because I wanted a professional qualification, but also because there are lots of interesting jobs related to nursing that don't necessarily involve working on a hospital ward. The research side of health has always interested me and I wanted to have more background knowledge of health and health care.'

Genetic counselling

Christine went on to work as a research assistant on a drug trial and then as a genetic counsellor. 'There are quite a lot of interesting research projects that utilize nurses as a part of the team,' explains Christine. 'When I decided to do my nurse training I wanted to become a genetic counsellor.' Although she enjoyed the work she found that there are only a handful of genetic counselling units in the country and so career opportunities are limited. 'It's a small world and very hierarchical. I don't like rigid structures and hierarchies.'

Moving to work in a collective suits her well. 'Nurses were traditionally trained straight from school and not taught to think for themselves, to be critical or to be interested in research. This is changing now but even so, for many nurses this way of working would be an alien concept.' Yet as attitudes change, so more nurse practitioners in conventional GPs' surgeries are gaining the autonomy they need to provide a good service for their patients. 'I could do research or audit type work here as well, such as discovering patients' attitudes towards various issues, or additional services they would like us to offer. If I had more time I could do this but it's very busy at the practice, and when I'm here I tend to work flat out.'

Is this the career for you?

'You need to enjoy working with people to do this job. When you're on the reception desk patients can be very demanding but most of them are extremely nice and it's rewarding because most of them are very appreciative,' explains Christine.

'You have to like listening to people too,' she continues. 'And try to help them find solutions to their problem in changes to their lifestyle. You have to be analytical in your thinking and be prepared to learn new things and add to your knowledge base all the time.'

As well as all this, and an ability to cope under pressure, you need to be a thinker: 'It can be intellectually demanding because people's problems are not always straightforward and you do need to know a lot about health and how people's bodies work.' As an example Christine cites the woman who was having problems with pre-menstrual tension (PMT). The patient had tried a medication that worked at first but then didn't help. 'We talked about her lifestyle,' relates Christine. 'Diet plays a role in PMT so we started talking about that. Research shows that it's important to eat regularly, but this patient never ate breakfast. We discussed the importance of eating regular high-carbohydrate snacks.

You've got to be prepared to listen and to give patients time. I also think that we can't look at health and wellbeing in isolation, they are often related to other things going on in our lives. This holistic approach requires you to be broad-minded and non-judgemental.'

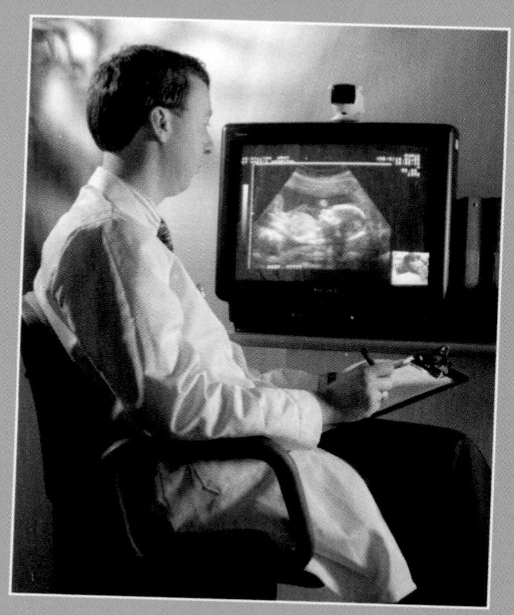

New technology is changing the way medicine works. Not many GPs' surgeries have long-distance foetal scanning equipment yet, but when they do it can help save lives, particulary in remote areas.

Career planning

Making Career Connections

Talk to staff at your GP's surgery and find out about the wide range of career options in community medicine.

Contact the Industrial Common Ownership Movement (ICOM) in Leeds LS1 6DE for information about working in a collective.

Talk to your careers adviser about careers in health work. Look at the different types of work and consider the different qualities and skills (personal as well as academic) needed for these.

Write to the Royal College of Nurses to find out how to train as a nurse. Your school or local authority careers service will be able to help you.

Getting started

Interested in being a women's health worker or nurse practitioner? Here's what you can do now.

1. Study sciences, especially biological sciences, at school and college.
2. Watch the staff at your local surgery or hospital next time you visit. See what each member of the organization does.
3. Do your own research projects on medical issues that interest you. Use your library and the Internet, and ask for help from your librarian or anyone you know working in medicine.
4. Enrol on a first aid course or join the St John Ambulance Cadets.

Related careers

Here are some related careers you may want to look into.

Paramedic
Drives an ambulance to an accident or emergency situation and uses life-saving equipment, sets up drips and administers certain drugs. Takes the injured or ill to hospital.

Doctor
Diagnoses and treats a wide range of medical conditions, or specializes to work in one particular area of medicine. Can work in surgeries, hospitals, clinics, workplaces and in the community.

Medical social worker
Works alongside medically trained staff to arrange care for people whose lives are disrupted by medical problems. Arranges practical or counselling help, or provides information.

Nursing auxiliary
Assists qualified nurses with day-to-day tasks, usually in hospital wards. This role will develop under it's new title 'health care assistant' to allow more training and responsibility.

Future watch

There will always be a need for committed, professional people working in health care. As the National Health Service develops, so more interesting roles with additional responsibilities are opening up for qualified nurses.

Some doctors prefer to see every patient themselves, as the GP is legally responsible for a patient's care, but this is seen as an increasingly old-fashioned view. The number of nursing practitioners is likely to increase as doctors have ever-increasing workloads because of rising patient expectations.

There are plenty of other types of health care careers. These speech therapists are working with children to ensure that their use of language develops normally.

Terry Clark

Systems Engineer

PERSONAL PROFILE

Career: Systems engineer. 'My work is part of information technology; I work in the testing side rather than development. This really means I have to see if new software acts in the way it's supposed to.'

Interests: 'I like football, chess and swimming. My German Shepherd dog is one of my main interests. And I have a busy social life – that's what it's all about.'

Latest accomplishment: 'I got two achievement awards for work I was doing as part of the testing team on a recent project.'

Why I do what I do: 'I went to a special school for disabled children. My motivation was to say, 'no, I'm not going to be a drain on the state, I'm going to look after myself. I enjoy my work – if I didn't I'd do something about it.'

I am: 'Methodical, very good at conceptual understanding, which means having a logical view of what the world is about. I'm good at thinking, analysing and telling people what I want – I'm a good team player, a social animal.'

What I wanted to be when I was at school: 'A shopkeeper: I had no conception that I would have to buy a lease and stand there for 14 hours a day for a pittance. We just didn't know about computers then.'

What a systems engineer does

Systems engineers design, build and test systems for customers with a particular need. Although systems engineers can work in other disciplines they are usually associated with computer software. 'The bulk of systems work is either development or testing, though there are also designers and support staff,' explains Terry Clark, a senior systems engineer employed by British Telecom (BT). 'I'm a verification, validation and testing (VV&T) manager,' explains Terry. The whole of the systems team works together to provide customers with computer software systems that meet their needs at a price they're prepared to pay. 'The customer is usually happy although they may want some changes made. We don't deliver it all at once. We drip-feed the systems over a period of time and while we're working on stage two we get feedback on stage one so we can make any necessary changes.'

The need for computer systems

'All our systems engineers work for internal BT customers,' explains Terry. This means that they design and build software systems to help other parts of the organization operate as effectively as possible. These other divisions pay for the work from Terry's division, so although they are all part of the same organization, they act as separate customers and must be treated as such. 'The total software engineering division in BT is probably the biggest in Europe,' says Terry. 'We provide the systems that BT needs to sell and promote its services to customers.'

Technology allows people to work from anywhere in the world – although most of them choose to stay indoors. This teleworker is 'plumbed in' to a computer network so that he can work anywhere he wishes.

A BT engineer designed this 'talking head' to help deaf people understand 'telephone' calls. The face moves to allow the user to lip-read.

Hardware: software

Computers can be divided into two main parts: the first is the physical boxes, screens and so on, which is the 'hardware'. The rest of the computer consists of the 'software' that is loaded into it, which comes in the form of computer programs, the coded instructions that tell the computer how to operate. Some people do work in computer hardware making the physical items, but the vast majority of computer workers are involved on the software side.

Reverse engineering

Standard 'forward' engineering begins with an idea for a system somebody wants. A specification is written, the code is devised and it is tested to make sure it works.

With reverse engineering there is no initial design or set of requirements. Instead, the programmer works with the code of an existing but out-of-date system. It could contain 100,000 lines of code and there might be no one from the original development team to help understand it. The software is analysed to see if it can be moved back to the conceptual level and changed to reflect present needs, thus completely changing the system without starting from scratch. This is done with a process called impact analysis and is a niche area of expertise.

Research

'As well as all that development and testing,' adds Terry, 'are designers and people looking to the future with all sorts of policy making.' Terry was himself involved in this part of the business for ten years. 'I was looking at the tools, procedures, ideas and methodologies that would benefit BT in future years. It was great fun to go off and investigate things in terms of mathematical statements and so on. There was a wide variety of work, and all looking ahead to something different.'

Now Terry is involved in systems testing. 'You've got to apply all the same principles to whatever the end goal is,' he explains, 'and try to do your best within the company pressures. I'd like to think that I would give any job total commitment.

All in a day's work

Terry may work in one of two locations: where his team is working, or in the main VV&T centre, where his bosses and many other systems people are located. His work is project based, and when a project is completed, the team is disbanded and people form into new teams for new projects. 'This is a flexible use of resources,' explains Terry. 'People can either treat it as an opportunity for learning new skills or as a disruption, but it does lead to a more flexible and skilled workforce among those who cope well with change.' Terry's current team numbers six people whom he manages directly, and another indirectly. The main development team is based in the other office so this is where Terry's testing team is also based.

'I have to manage my team and all their interactions with the development team,' says Terry. 'I have a good conceptual idea of what goes on in both teams.' The individuals are doing specialized bits of work, and Terry will discuss any problems with them. 'I'm employed as a manager, not a doer,' says Terry. 'My role is understanding, facilitating, giving advice, general direction and leadership. I have to keep all my people happy.' He does this by keeping in touch using the telephone and e-mail, as well as visiting when necessary. 'I like to speak to every team member at least once every couple of days. I could sit back and say that as nobody has rung me it must be hunky dory, but if they don't phone me, I phone them. There must be trust both ways.'

Terry is based in open-plan offices, and uses the most up-to-date technology to keep in contact with his staff and other colleagues. There are times however, when nothing can take the place of a face-to-face conversation.

Overcoming disability

Terry suffers from cerebral palsy and is also severely deaf. 'I have a lack of co-ordination in all four limbs,' explains Terry, 'I walk reasonably well but I do have problems carrying things like plates of food. I can sign my name but writing makes me very tired. When I came into BT and wanted to be a programmer I was given a screen and keyboard and was more than equal with most of my contemporaries. It's a piece of equipment that made me come from 60-70 per cent efficiency to 110 per cent.'

Terry's deafness causes him more problems than the cerebral palsy, as many people are unaware of it. 'Meetings are a nightmare,' he explains. By the time he has located the person who is speaking and started to lip-read, someone else may have interrupted, and Terry has to track this speaker and lip-read there too. But other activities are not too difficult. 'I push the telephone hard against my ear so that the noise goes through the skull rather than the ear drum,' he explains. 'In most things, I get by. Most people recognize that I have this limitation, and I try to compensate in other areas.'

Thinking time

'One of the reasons I spend time in this office,' says Terry, 'is that I get sidled off to meetings where I can help other teams to plan their work. My boss uses me this way because my skills lie in thinking and conceptualizing and I

Left: Because Terry is severely deaf, he has to press the telephone hard against the bone behind his ear to hear what is being said. Sometimes he will ask a colleague to double-check a telephone number that he has been given.

Below: A BT engineer designed this 'talking head' to help deaf people understand 'telephone' calls. The face moves to allow the user to lip-read.

sometimes think of things that would otherwise get overlooked.' Terry explains that he doesn't miss the technical work he used to do: 'I had many years in the business at a time when the technical side was exhilarating and fun, but once you've done it, you know you can do it and it's no longer fun. Whereas the ability to get the whole picture, and using that ability to forward projects is always going to be new and exciting.'

Activity

Building a system

'Systems' need not mean computers. For example, a systems engineer might design and build a new means of organizing an office's or school's incoming post; or plan the electrical circuitry for a new stage lighting system or a new one-way system around the town.

You will need:

A project to work on, such as one of the above, or a new idea for your home, school or another environment, or indeed a computer system. This may be a real project or a 'virtual' project where you plan and implement – but only on paper.

Possibly a team to work on it together, although you can do this alone.

Procedure

1. If you have a team, once you know the work that is involved, decide who will perform which functions.
2. Decide who the project is being designed for (the client).
3. Look at the old system, ask questions about it, take notes, then draw a flow diagram of how it works.
4. Ask the client exactly what they need from the new system. Write this down and draw another flow diagram. This is your 'specification'.
5. Go and talk to your client about your proposed changes. Ask whether they think they are appropriate. Do they seem practical, cost-effective and easy to use?
6. Listen to their comments and change your plans accordingly.
7. Start work on the new system: find the components you need and put them together in a clear space. Test the system. Does it work? Ask the client to look at it. Is the client happy?
8. Keep going back to the original idea and the working plan. Are you keeping an objective view over all that is going on? Is your work going to schedule? Are costs being exceeded?
9. Deliver and install your client's new system. Does it work? Should you dismantle the old system or keep them running in tandem to begin with? Is the client still happy? If not, why not? What can you do about it?
10. Review your work and everyone's role in the team. Were some people's input more important in certain areas than others'?

How to become a systems engineer

Terry always enjoyed maths and physics: 'But there was no one at my school to teach these at 'A' level,' says Terry, 'So I did two maths 'A' levels off my own bat. I had the motivation that I knew I could do it, and I did. I was confident in my own ability, big-headed and over-confident too perhaps. But I have to be, it's part of my own survival mechanism, and becomes more so as time goes on.'

He followed this with an HND before joining BT as a trainee. 'I wanted to be a programmer. The thought of getting a machine to do something by giving it instructions was quite novel in those days – the thought of being in control. It was an up-and-coming technology and I thought I'd be quite good at it – and I was.' During his first few years of work he completed a postgraduate diploma in business studies, and from then moved around and up within BT.

Now that computer technology is commonplace, there are far more openings for trainee programmers, analysts and systems engineers. Most training schemes require a degree or HND to join, though not necessarily in a relevant subject. Many organizations use a form of testing to check candidates' logical thinking and basic numeracy.

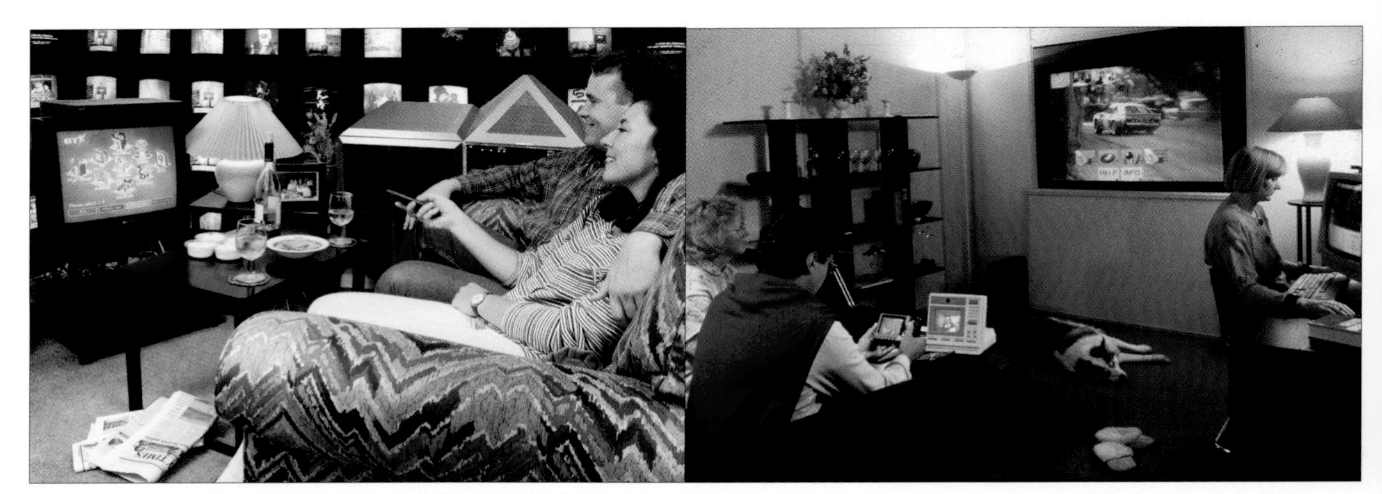

How long will it be before our homes look like this? Perhaps not too long. In fact, some people already spend their days and nights surrounded by state-of-the-art technology.

Is this the career for you?

Technology is changing so rapidly that anyone entering this area of work needs a firm commitment. In areas such as accountancy and law, the basic principles remain in place for many years, perhaps even centuries. This is not the case with computer technology: 'Today's technology will be obsolete by this time next year. You need to be able to cope with rapid change, to be committed. Technology is innovative – you have to keep up with it. Timescales can be rather arbitrary and we're always pushing deadlines,' explains Terry.

'You also need a sense of humour,' he continues, 'you can't take everything seriously. There's a lot to be said for good teamwork and trust in the people you work with – it makes it all more worthwhile. The team is important, we've all got to pull together and all move in the same direction.

For his management role, Terry says there are some basic rules that he works with: 'If there are serious problems I escalate them through the management tree. If they are just squabbles I sort them out. It's about encouragement, sympathy, cajoling, telling off, all the normal management techniques, including carrots and sticks to get the job done.' These are management skills learnt in the workplace but they do require a basic understanding of people and what motivates them. And is it all worth it? 'I love work,' says Terry. 'Though I'd rather have the Chairman's job,' he laughs.

Career planning

Making Career Connections

Contact the local office of a telecommunications company. Ask their personnel department for information about careers in software engineering.

Approach your careers adviser for course and entry details for likely sounding degree and HND programmes. Write to the universities concerned for their prospectuses.

Use your contacts to find a local office that uses up-to-date computer technology. Ask whether you can visit to see this technology in action.

Invite a higher education teacher into your school to speak about courses in computer studies. Ask how important this is as a basis for a career in computing, or whether a general degree is equally useful.

Getting started

1. Keep studying a broad range of subjects. Remember that communication skills are as important as the technical skills you will learn later on.
2. Ensure that your keyboard skills are excellent. Enrol on a private course of evening or weekend classes to develop these if necessary.
3. Read the specialist press for news of developments in the information technology field. What repercussions might these changes have on jobs in the longer term?
4. Look at newspaper advertisements for jobs in the systems engineering field. Note the many different terms used to describe the different functions, and sometimes used to describe the same function.

Related careers

Systems analyst

Works closely with clients and programmers to design software for specific needs. Closely related to, and often synonymous with, systems engineer.

Computer programmer

Writes the lines of code that form a program, according to specifications laid down by the designer. May specialize in one particular computer language or type of machine, or an area such as banking systems or leisure software.

Computer support engineer

Installs and maintains the hardware necessary for analysts and software engineers to do their jobs. May be involved in upgrading systems and installing networks.

Future watch

'There will always be a need for systems engineers,' says Terry. 'What will change is the technology and the sort of systems that people want. This industry is young – only 35 to 40 years old. It's growing and will be here forever. When we send a rocket to Saturn, computer programs will control it. The boundaries are limitless.'

Amita Jariwala

Science Teacher

PERSONAL PROFILE

Career: Science teacher. 'I specialize in chemistry. I am also a Key Stage 3 co-ordinator.'

Interests: 'Sport – I watch football and rugby and take part in lots of sports, particularly tennis and aerobics. I enjoy walking. I'm usually out at the theatre or cinema with friends. I'm not often in.'

Latest accomplishment: 'I've just been promoted to the Key Stage 3 science co-ordinator after three years teaching. It was the job I wanted to do – I'm very pleased.'

Why I do what I do: 'I wanted to work with people and I like communicating information. I find that an easy thing to do.'

I am: 'Very independent, organized, bubbly and a good communicator.'

What I wanted to be when I was at school: 'A doctor at first, then I thought about being a teacher.'

What a science teacher does

It's easy to assume that teachers just turn up to lessons, talk a bit and then go home. But ask any teacher and they'll tell you that this is not so. Amita Jariwala is a science teacher at a large comprehensive school. Her school has 1500 pupils and 12 science teachers. She teaches all the sciences and all ages of students, but she specializes in chemistry. 'Teaching is about communicating with other teachers as well as the children,' she says. 'It's a two-way thing, an information exchange.' So although Amita has a set list of information and methods she needs to help the children learn (the National Curriculum), she also needs to listen and obtain feedback all the time.

Amita teaches a year 10 practical lesson. The students are looking at metal reactivity. She ensures students wear their safety glasses at all times when using chemicals, and that they dispose of chemicals safely at the end of the class.

'We have schemes of work,' says Amita, 'so that every science teacher in the school will be doing almost the same work. All the students in any years will be taught the same thing, though in a different order.' The science teachers contribute to this by writing lesson modules, which are then circulated for comments from the other teachers. 'These constantly change,' says Amita. 'If we find a new practical project to do, for instance, we'll incorporate it.' Amita then pulls together all these modules in her role as Key Stage 3 Co-ordinator. All this is before seeing a single pupil!

Airport baggage

Other parts of her work include organizing extra-curricular activities. These can take a large amount of teaching time, requiring other teachers to take her lessons, but she feels it is worth it for the extra dimension they give the students. 'I am helping to organize a group of sixth form girls to go on a Women in Science and Technology course at a nearby airport,' she explains. 'Then they go on to the Technical College and build a baggage transit system. I need to sort out the minibus and all the paperwork, then drive the minibus on the day.'

Even within her teaching time, which is timetabled to cover every lesson except four in a week, she needs varying approaches for different groups of students. 'I taught two lessons on atomic structure yesterday,' she explains. 'The year 12 ('A' level) groups were doing the arrangement of electrons in orbitals. There are only five in the group and they all chose to study it. They all understood it and were interested. Then I taught a year 10 (GCSE level) group on the same subject. It had to be taught in a very different way. The language we use is different too. They really got into it and were responding to all the questions I asked. They were happy at the end of the lesson that they had actually understood.'

All in a day's work

A typical school day starts at 8.15 for Amita. She goes to her staff room to check her pigeon hole and the cover board. 'I have a free period today,' she explains, 'so I check to see if I have to cover for another teacher. It could be any lesson. I'll usually do some marking but if it's a science lesson I'll sometimes get involved. I've had to help children with French worksheets,' laughs Amita. 'My French was better than theirs.'

Although the students move on at the end of each lesson, Amita stays put. This is 'her' laboratory and most of her lessons take place here. She has a few minutes to prepare before the next class arrives.

After checking the board she spends time with her form group. 'As well as taking the register I deal with any problems and check and sign homework diaries. I give out award slips and absence slips.' Then she goes to her first lesson. Most of her lessons are in the laboratory, even the theoretical ones, though the students sometimes need to use the computers in other rooms. 'For a practical session, the equipment is brought into the laboratories by the technicians. I need to organize how the children move around the equipment so that it is safe for them. This year 11 class is doing an investigation into rates of reaction of sodium thiosulphate with hydrochloric acid, and how it's affected by temperature. I go around making sure they're doing it all properly and safely, and that they're doing it at all. I have to really keep some of them at it. They wrote the method themselves and predictions of their results in a previous lesson. Next week they will work out their conclusions and draw their graphs. One hour is not actually enough time once we've talked about what we will do, got the equipment out and set up, done the practical, allowed ten minutes for packing up and then talked about next week. The pupils have to take responsibility for returning the equipment at the end of the lesson. The laboratory should be left in exactly the same state as when they came in.'

Senior Laboratory Technician, John Shirley, removes the chemicals needed for the next lesson from the cupboard. All chemicals are kept safely under lock and key; so too are the laboratories when they are not in use.

When Amita is on break duty she doesn't have a mid-morning rest. She teaches two classes, supervises the pupils' break period, then goes on to teach another class before lunchtime.

At the end of the day is another registration session. After the pupils have gone home, Amita sorts out the lessons for next week, when she will be away on her trip with the year 12 students, before she too goes home. 'I might need to stay late for staff , curriculum or year team meetings, or for an INSET training session, sometimes for parents' evening. I need to prepare ahead for these. I must have an idea of what I am going to say to the parents. They need direct information about what their children should be doing, or that the child is actually

Prior experience

Many teachers leave university and feel they want some experience of work outside education before they come into teaching. This is what Amita did. 'I just wanted to get out of the system for a while,' says Amita. 'I wanted to see what the working world was like. I also think that once you've done a certain number of years in education, you need a break.'

Amita spent several years working for a pharmaceutical company, visiting doctors and explaining the advantages of certain drugs for their patients. This job is known as a medical 'rep'. 'You have to be trained within your area better than the doctors are,' Amita explains. 'You need to be able to answer every question, and also to know how to sell. I enjoyed the training and doing the job for the first few years. I liked using my degree, the freedom, and the buzz from reaching my sales targets.'

The pharmaceutical industry is huge and invests heavily in research, so products are always changing. Some reps stay in this work for a long time, others, like Amita, move into other areas. There are many other jobs for scientists in all areas of the pharmaceutical industry, such as in the creation of new drugs in research laboratories.

doing very well. You can't just say bad things, it's nice to give them some positive feedback. In a lot of cases it's all very good, but I do have to let them know where they need to improve.'

Amita does as much of her marking as possible during school hours. 'A set of books can take an hour and a half. I sometimes spend my lunchtimes marking. The children see how well organized you are by how often and well you mark their books.'

It's a fact

Teachers don't really have thirteen weeks' holiday a year! 'I have work to do over half terms and the Christmas and Easter holidays,' Amita explains, 'because I have coursework to mark. At the end of the year I do as much preparation as possible for the following year before I start the summer holidays. But the long holiday really is an advantage. The teachers are all as tired as the pupils.'

This scientist works in pharmaceutical research, developing new drugs for prescription and over-the-counter sales.

Activity

Teaching science

People carry on learning throughout their lives. Imagine how much new science will be discovered once you have left education, then think about people who left school in, for example, the 1950s. Have they kept up with the basics of the new science?

You will need:

- some recently discovered science
- an older person to teach it to
- whatever teaching resources you think will be useful.

Procedure:

1. Find older people who are willing to learn something new about science. Ask when they left education.
2. Find a 'piece' of science that has been discovered since. An example might be Crick and Watson's discovery of the double helix structure of DNA, which revolutionized the way we see living things, or the discovery of the electron. Ask your science teacher for help with this. Ask your 'pupils' what they already know about this piece of science. Some people stay in touch with new discoveries while others don't.
3. Think about how you will teach this science to your 'pupils'. Consider:
   - their attention span
   - their current scientific knowledge
   - their interest in the subject
   - whether they enjoy using diagrams or written or spoken words
   - if there is a practical way of showing it, such as demonstrating static electricity by sticking balloons to walls after rubbing them.
4. When you have finished teaching this piece of science, ask your pupils to recap for you, so you can tell whether they have understood. If not, why? How could you improve your teaching methods?
5. Remember to give a handout summarizing the main points at the end of the lesson.
6. Ask whether your pupils are interested in learning more another time.

How to become a science teacher

'I did chemistry, biology and maths at school,' says Amita, 'then a degree in chemistry and from there went straight into being a medical rep. I always liked chemistry. I found it easy until 'A' level when I found how difficult it can be: all the concepts are very abstract. Science is the most abstract of subjects because you can't see it. You really have to sit down and picture it for yourself. But the degree itself was easy – I sailed through that.'

After several years as a medical rep Amita started training other reps: 'I enjoyed that a lot. I did think about staying in pharmaceutical training but decided I should go back to doing what I really wanted to do. I did a postgraduate certificate of education (PGCE) which I loved. It was exciting, with new people and lots of things to learn.' Amita was asked to apply for a job by her current school. 'My first year of teaching was very difficult,' she admits. 'There was lots of planning and also problems as one of the other science teachers was very ill so I was doing some of his job. It was hard, but it was good fun. I still have days when it's totally frustrating and I just think "why am I doing this?", but most of the time I go in happily. It's different every year with new classes, different pupils and responses, new personalities. It's great.'

During a free period, Amita catches up on paperwork at her desk in the prep room.

Is this the career for you?

'You have to be very patient,' explains Amita, 'very organized, have good communications skills and know how to talk to the children – and not the same way to every child. You know this by listening to them. You also need very good discipline skills: you have to have a "presence". You can learn tricks, but with classroom management, you either have it or you don't. If you're halfway there you can learn it. I knew that I could do it, that I can get people to listen to me.'

Communication is essential. 'You need the ability to talk to people at the right level. You can't be isolated from the other teachers. You need to pass on information about the children. You have to have a sense of humour all the time. When you walk into a class they don't want you to look grumpy. If you are smiling when you walk in they'll do what you want. It is important to have good acting skills. Once you're in the classroom you've got to be yourself but it's an act as well. You need a lot of energy but it's a problem if you have too much energy – you need a nature that's calm when necessary.

Every teacher must work hard at keeping up with new developments in their subjects. 'It is important to read the journals and watch programmes about science on television. You've got to enjoy science and be enthusiastic about it: if you are, the children will be." Amita herself is full of enthusiasm for her work: ' There's always something new to do – it's very creative.'

These year 4 children are learning basic scientific experimentation techniques. They discuss how their senses can help them identify white powders; they also learn that it is not always a good idea to smell or taste in science!

Career planning

Making Career Connections

Speak to your science teachers about their work. Find out what is involved in addition to classroom teaching.

Organize a group visit to a science lecture at a nearby university.

Ask your careers adviser about teacher training. Look at the different options available and send for prospectuses for the courses that interest you. Investigate other teaching options: primary; further and higher education; special needs; training within commerce and industry; adult education, and so on. Ask your careers adviser for help.

Contact a nearby university department of education and see whether a tutor can come to speak to your group about teaching methods. Ask whether these are changing, and how.

Getting started

Interested in being a science teacher? Here's what you can do now.

1. Concentrate on your studies, especially all three sciences.
2. Add to your knowledge base by reading more about science: look in your library or on the Internet for the science journals. Keep a look-out for television and radio programmes about science.
3. Join or set up a science group at your school. Look at new topics each week and invite speakers on areas that interest you.
4. Ask to help out in science practicals or lesson preparation at school. See how the teacher helps the students learn.

Related careers

Here are some related careers you may want to look into.

Training officer
Works with adults teaching them how to do their job well. Organizes, writes and delivers courses, assesses participants, liaises with managers regarding training needs.

Researcher/manager
Works in the research side of large industrial organizations; trains in scientific research and management techniques.

University lecturer
Combines research with teaching undergraduate and postgraduate students, also supervises research projects.

Explainer
Works in a museum explaining science exhibits to interested people.

Future watch

'Science is a core subject,' says Amita. 'There's always a shortage of science teachers, especially women teaching chemistry and physics. You can almost guarantee that you will get a job at the end of your training.' Amita's career also demonstrates career progression for good teachers, with additional responsibilities available for those who want them. Teaching science and technology is also increasingly important among adult learners who need to keep up-to-date with new advances. The changing nature of science is another reason for a good future in teaching: 'It's very interesting and always stretches you,' explains Amita. 'I don't get bored.'

Phil Smith

Biologist

PERSONAL PROFILE

Career: Biologist. 'I collect and analyse water samples – these come from private ponds that have gone a funny colour, or from rivers.'

Interests: 'I'm heavily into hill walking and natural history, which ties in with walking. I like sport: I play golf and football.'

Latest accomplishment: 'I've been appointed as the biology team leader for part of this area.'

Why I do what I do: 'Because I care about and want to improve the quality of the environment. I'm interested in ecology and living creatures; why they occur, where they occur and how they respond to whatever's going on in the world.'

I am: 'Hardworking, a bit of a perfectionist (scientists tend to be). I have fairly clear ideas of where the group should be going and what we should be doing.'

Who I wanted to be when I was at school: 'Jacques Cousteau. He was a marine biologist who brought the undersea world into the living room with his television programmes. That's what inspired me to do this but there were no jobs in marine biology so I went into the freshwater side.'

What a biologist does

Biologists can work with any type of living thing in the animal or plant kingdoms. Most of the work available outside research departments is concerned with our environment. Phil Smith works as a biologist for the Environment Agency, an organization set up by the government to manage and police the environment. Most of its work is concerned with water and keeping our rivers and lakes clean, as polluted water has a more immediate effect on us than polluted land, although this emphasis is now shifting.

Phil uses a standard method of collecting a sample from a stream or river. It is called 'kick sampling': he stands upstream of the pond net and kicks the stream bed with his foot for three minutes. He moves around to the different areas of the stream (weed, gravel, margins. etc) during this time to sample the areas roughly in the right proportions. He then empties the sample into a bucket and pours a preservative over it so that it will keep for analysis in the winter. If there has been a specific pollution incident he will keep the sample live and analyse it immediately as this makes better evidence for a court prosecution. Phil is sure always to wear his life jacket and only go near the river when accompanied. These are basic safety precautions.

Natural and man-made stresses

Phil's work is concerned with monitoring 'stress' levels in the waterways in his area: stress can mean pollutants finding their way into the water or other effects such as low rainfall causing rivers to dry up. 'We're really here to monitor the direct effects these changes are having on the environment,' says Phil. 'If someone tips pollutants into a river, such as a farmer allowing slurry to escape, you can detect this by taking water samples. But if you want to know the impact this has had on the environment you have to look at the creatures in that environment.' When pollution occurs, the water itself may only be affected for a short while, whereas the creatures in the water will paint a picture of what happened for a long time afterwards. This means that pollution can be detected even after the initial effects have worn off.

How can tiny creatures measure water pollution?

River invertebrates live in the mud on the river bed and don't move very far. Biologists look at the type and number of macro invertebrates in a part of a river to see what has been happening to the water.

Mayfly, stone fly and caddis fly larvae are very sensitive to pollution. If there are none or very few in a river bed sample, it indicates that the water is or has been polluted. Other creatures, such as midge fly larvae and pond snails, thrive in water polluted organically with substances such as sewage or farm fertilisers. Phil and his colleagues take samples at different points, perhaps above and below a waste pipe, and pinpoint the entry of pollutants into the river in this way.

Other creatures, such as fish, also show the effects of pollution but they often swim great distances, so are not as helpful a guide to what's happening at any particular spot on the river.

Future generations

'We are also concerned with sustainability,' says Phil. 'So that the environment we are passing on to the next generation is no worse than the environment that we inherited, and where possible, is improved.'

'As well as analysing changes to the creatures in the rivers, the plants are sampled and analysed to give an indication of how successful a dredging programme has been: 'Dredging is part of river maintenance,' says Phil. 'The river must not be so choked with vegetation that it will flood property during heavy rainfall. Our engineers dredge the rivers but there is the potential to do a lot of damage so we look at the site first and draw up a work plan that will cause minimal damage to the environment.' The changes can later be assessed by sampling the vegetation around the dredged site.

All in a day's work

Phil mainly does two types of work: sampling and analysis. 'It can take anything from one hour to two days to analyse a river sample,' says Phil, 'although one day is typical.' Phil starts work at 8.30am and can be looking for creatures in his sorting tray for the rest of the day. 'Doing analysis means you're staring at trays day in and day out. You have to generate your own interest out of it through interpreting what's there and what it can mean.' Although there are certain score systems and computer models used in the process, the real interpretation is down to the skill and experience of the biologist.

Phil washes preservative out of a sample in the fume cupboard. The chemical used is formalin, a potentially harmful irritant, so he wears rubber gloves and safety glasses.

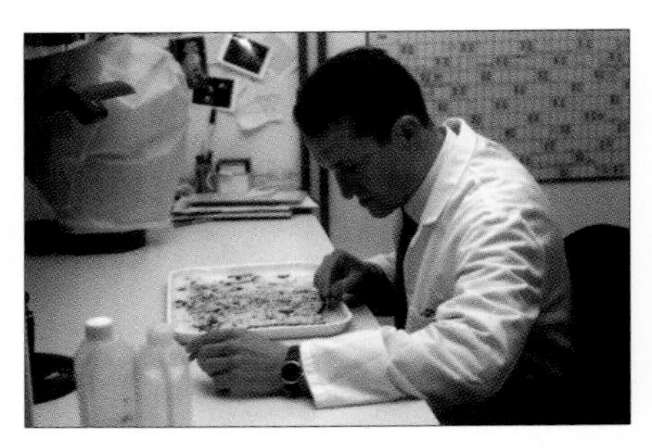

The individual invertebrates are taken from the sample and placed in a spot tray for counting.

Creatures score points

To analyse a sample Phil divides it into different sized fractions using a stack of three sieves. This separates the animals from other material. 'I then put a small amount from each sieve in turn into the tray and pick out the little invertebrates. The macro invertebrates tend to be in the middle-sized sieve with the 1–3mm mesh.' Phil looks through the trays picking the animals out, identifies them under a microscope and counts them. This leaves him with a long list of animals. 'I have to translate this list for non-biologists,' Phil explains. There is a scoring system whereby creatures that are sensitive to pollution score high and those that are not get low scores. 'This information is fed into a computer which adds data such as the depth and width of the river at that site, the composition of the river bed, the rate of flow on that particular day, and so on. I then compare what I have found with what the model says I should find and this gives a good idea of the conditions at the site. This model gives a scientific basis for deciding something is wrong; it's a powerful tool in convincing people that there is a problem.'

What is an invertebrate?

An invertebrate is an animal without a backbone. Phil and his colleagues look at macro invertebrates, greater than 1mm in size. These have been researched and used as a measure of water quality over generations, so now biologists understand what it means to find certain creatures in the water.

Incident alert

Phil is always ready to be called out to an incident. 'A water quality officer may ring at any time, possibly alerted by a member of the public who has seen dead fish in a river,' explains Phil. 'The officer will have done an initial check, maybe taking water samples for chemical analysis. 'Then, to find out how much damage has been caused, Phil is called out. He collects all his sampling gear: nets, buckets, waders and lifejacket, and goes to meet the officer on site. 'Then I design a survey,' explains Phil, 'to best measure the incident's impact. I need to consider other influences such as tributaries and streams, or other pipes, and could well have to defend my sampling methods in court. I have to think of the best way to survey and then carry it out immediately because time is a limiting factor. I take samples up and downstream of the pollution source, and further samples at set

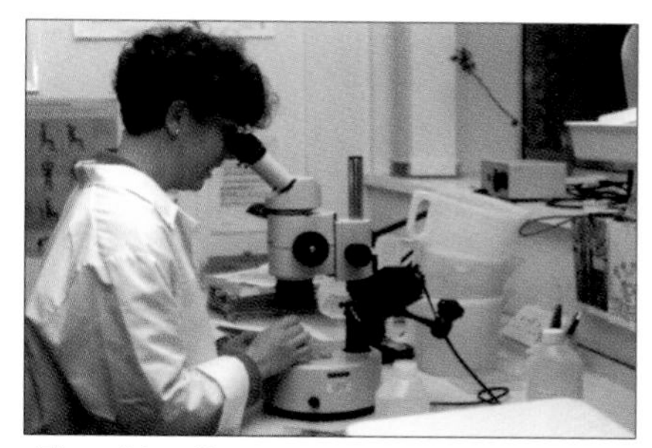

Elinor identifies macro invertebrates thorough a low-power microscope. She checks them against a key and writes her findings on a score system sheet.

distances, typically up to six samples.'

Complete evidence

'I put the whole sample into a tray on the riverbank and record how many invertebrates are alive and dead,' continues Phil. This is different from the routine sampling technique where the whole sample would be killed by preservative before being taken for counting: with a pollution incident the data needs to be as complete as possible. Phil takes the samples back to the lab where the biology team is waiting. 'It's all hands to the pump,' says Phil. 'We check the samples again. Some invertebrates may already have died but we need to be sure we haven't missed anything. Our results can provide very powerful courtroom evidence. It's very applied work and very relevant. An unusual aspect of this job is that I do it right from planning through to writing the report, from beginning to end.'

Seasonal work

In the winter, river levels may be too high to obtain samples, and plants die back and are less easily identified. Most routine sampling takes place in the summer, with the samples preserved for winter analysis.

Activity

Pond dipping

Ponds and rivers contain a huge variety of wildlife and it can be fun and interesting to look at what's there and what it means.

You will need

- a responsible adult to go with you
- lifejacket and wellington boots
- sampling equipment: net, bucket, tray, forceps or tweezers, other small pots
- information to help identify what you find
- a stream or pond.

(Safety note: never go near or enter water alone, however safe it looks. Even with an adult present, always stay in areas where you will be able to call for help. Always wear a lifejacket, and rubber gloves to avoid disease. Always wash your hands after touching river water and samples.)

Procedure:

1. Look at your site carefully. Design a sampling survey to cover different parts of the stream.
2. Take samples by holding the net downstream of your foot and disturbing the stream bed. Take an 'average' sample including each of the different types of stream bed.
3. Tip your sample into a white tray and pick out the creatures. Place different types in different pots and count them when you have finished. Alternatively, take the sample away and analyse it later.
4. Use books and any expert advice to identify what you have found. The photographs on this page may help. You won't need a microscope for most types of macro invertebrates.
5. Compare what you find in one spot with what you find in another. How can you explain the differences? You may need to do more research to discover which animals like which conditions. Have you discovered any pollution? If so, where has it come from? And what can you do about it?

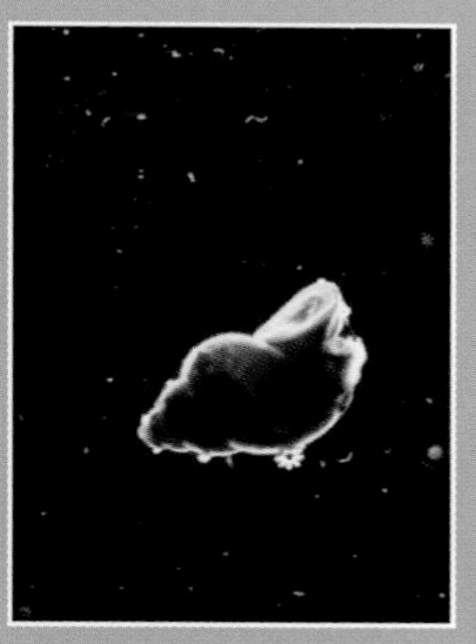

A type of pond snail.

Stone fly larva.

Damsel fly larva.

How to become a biologist

Phil has been gripped by biology since he was very young. His degree in biological sciences included studying freshwater pollution, which encouraged him towards his current job. 'It seemed to be a real application of biology in the real world, and there aren't many of those,' he says.

After his degree and an MSc, he worked in fisheries for a water authority for a while. 'That was quite different, it was all about team work, which was really fun, but it meant field work every day, often in muddy, dirty cold conditions, and it was hard work hauling in nets.'

He moved across to the biology side of the organization. 'It is a bit more science-based,' he explains, 'and a more complex job in a way.' He has worked as a biology technician and risen through various jobs to his new management role, due to start soon. 'A team leader still has hands-on work,' he says, 'but it is mostly management work.'

It is a very difficult area of work to get into and many biologists have postgraduate qualifications as well as first degrees. Phil explains: 'A lot of graduates want to get in, and there are very few opportunities.' Often, people are taken on temporary contracts to begin with, and those people will probably have had some work experience with the Agency first. This could be through a sandwich course that includes paid or unpaid placements. 'There's such a long training period,' explains Phil, 'that students are not much use for the first few months, but if someone is good we will try to keep them on as a temporary staff member.'

This mess of debris contains all the clues Phil needs to build a picture of the recent history of the water in the river.

Is this the career for you?

'You've got to be self-motivated,' says Phil. 'You've got to want to do it and to learn a lot from the job. You must take all opportunities to learn more and more – it's not something you can ever know fully.' Phil also says that you have to generate your own interest in the work or it can become like a production line: you need to be able to concentrate well, to be able to analyse the samples for long periods in the winter.

You also have to like going out into the field, even when the weather is bad, and not mind sitting in the lab for long periods of time. As with all science, you need to be happy working with numbers and computers and be able to write reports on your findings. These reports could be used directly to improve the environment so it's important that they are accurate and clearly written.

'The day to day work can be tedious,' confesses Phil, 'but I know I am doing something important.

Career planning

Making Career Connections

Ask your careers adviser about degree courses which include freshwater biology. Send for the college prospectuses. Ask too about relevant masters degrees.

Ask a researcher at your local university to speak to you about research careers in this area. Ask what the differences are between research-based and environment-based work.

Contact your local Environment Agency office. See whether a biologist can come to talk to your group about the work and the qualifications needed.

See if you can visit a biology laboratory or accompany a biologist on a field trip.

Getting started

Interested in being a biologist? Here's what you can do now.

1. Keep studying a broad range of science and maths. Remember that you need writing and comprehension skills though, so don't drop the arts subjects.
2. Ensure that you understand computers and are happy using them. This could be a good time to do a keyboard skills course out of school.
3. See if there are local or school biology societies. Go along to meetings and broaden your knowledge of biology that way. You may make useful contacts.
4. Read the popular science magazines. Cruise the Internet for interesting news from the research journals.

Related careers

Here are some related careers you may want to look into.

Conservation officer

Carries out field surveys, usually on plant life. Deals with planning applications. Works on projects with community groups.

Pollution officer

Investigates pollution incidents, collects water samples for analysis, deals directly with polluters and the public.

Fisheries officer

Surveys fish populations, rescues fish from polluted waters. Checks rod licences of fishing enthusiasts.

Research biologist

Studies a relatively narrow subject area in great detail.

Future watch

'Quite a lot of biologists are looking at water; in the future the land environment will be a big issue,' says Phil. 'We have to go back and re-use old industrial sites, and this means clearing them up first. There will be a more holistic approach to environmental management in the future, requiring people with a broader biological background.'

Vivienne Parry – Science Broadcaster and Writer

What do we increasingly depend on, but is becoming more complex and difficult to understand? The answer is science. Somehow, we need to keep up-to-date with new discoveries and inventions through our lives. The communicators who help us do this include Vivienne Parry, a broadcaster and writer who specializes in biology. 'So much of what people do now depends on an understanding of science,' says Vivienne. 'There has never been a greater need for there to be science people who can tell others not just about the nuts and bolts, but about the wonders of science.'

Two careers in one

There are two main parts to Vivienne's work: filming for television programmes such as *Tomorrow's World*, and writing articles for newspapers and magazines. The first could take place anywhere but the second is usually in her study at home.

'I love doing television,' says Vivienne. 'It's like heaven on a stick as far as I'm concerned. It allows me to combine diversity with interesting subjects. It's a real adrenaline rush all the time.' The work is not always as glamorous as people think: 'It can be pretty bleak sometimes, and there is no trailer to retire into. We may be filming outside all day in the depths of winter.' The call sheet will usually tell Vivienne what she will need for the day's filming. 'I always worry about those calls that say "wellingtons would be advisable" or "wear protective clothing". I hated doing a film on a maggot farm. The smell was disgusting!'

We all watch television, listen to the radio and read books and newspapers. Vivienne is one of the communicators who present scientific information to us through these media in a way that we can understand.

Vivienne enjoys the writing side of her career 'but it's a solitary job and I'm terribly gregarious. I can only do a certain amount of writing before I get bored being by myself and have to go off and do something else.'

Vivienne's media career started with writing. After gaining a degree in zoology she joined a research charity as their national organizer. This involved running research programmes and big events, and making science understandable to donors. 'I absolutely loved it,' she says. 'I did a lot of radio as well, and was at home one day a week to write, which is how I started building up my writing career.'

Fitting the work jigsaw together

A great deal of Vivienne's work takes place behind the scenes: she has research to do, production meetings to attend, and often hours, or even days, spent travelling. A filming day usually involves an early start. 'A call is the time I'm expected to be in a particular place, anywhere in the country. It's often 8.30 in the morning.' This may mean travelling the night before. 'And it's often at very short notice,' continues Vivienne, 'because we are completely reliant on the availability of the people that are to be featured in the film. I often have to just drop everything and go.'

The only routine in Vivienne's life is to swim half a mile every morning. Physical fitness is very important in such a physically and mentally demanding job. 'You have to be fit to do a lot of filming,' says Vivienne.

If you haven't got the freelance mentality or can't cope with insecurity, this is not the work for you. But for the right person it's a terrific job: 'It's everything I ever wanted and more,' says Vivienne. 'It's exciting, it's interesting, it's absorbing and they pay me too! I really love it. It's just a high.'

Getting started

1. Write articles for school, college or local papers. Keep writing up your new ideas and sending them in – be persistent.
2. Watch television presenters and examine their techniques. Try to gain an understanding of how a television programme is put together.
3. Keep news clippings on subjects that interest you. Learn to keep a finger on the media pulse this way. Use these as a base for your articles.
4. Learn to ask people about what they do, whether they enjoy it and so on. Become used to asking and gaining information.

Chris Self – Consulting Structural Engineer

Chris is a chartered engineer. He uses drawings like this to plan, explain and execute the work on site. Chris's particular expertise is in areas where the land is contaminated.

Have you ever wondered how buildings stay standing? Could you design one that does? And could you also design it to look good, cost the right amount and be friendly to the environment? This is what structural engineers do: they design structures (eg, a road, dam, jetty or building), to cope with the forces that are exerted by nature and by use.

Showing the supports

Structural engineers work as part of a team with the client, architect and other engineers. 'Client relationships are very important,' stresses Chris Self, a structural engineer. 'In some buildings the client will want the structure hidden, in others a client wants it shown. We calculate the loads and forces and then compare those with the strength of the elements we are designing, such as the beams, columns and floors,' Chris explains. 'We make sure that these elements are strong enough to cope with the forces.' Working this out involves using mathematical equations and computer-generated charts and calculations.

Chris specializes in 'brownfield sites'; areas where the land is contaminated. His work includes sending soil samples to a laboratory for chemical analysis, then working out a cost-effective way of dealing with the contamination. 'We don't want to take contaminated material off the site,' explains Chris, 'as this is environmentally unfriendly and very expensive. We see if we can clean it on site, either chemically or with water or heat treatment. It's an important area of work for engineers now.' Chris also arranges other site tests to find out, for example, the geotechnical capabilities of the soil and the load it would be possible to put on the foundations.

Crushing concrete

Materials are generally pre-tested but some, like concrete, are made up on site and Chris needs to be sure they are strong enough. 'When we pour concrete we also fill special cube moulds then take these to a lab for crushing,' says Chris. If it's not strong enough further cubes would be tested, together with samples cut from the site concrete.

Engineers need to be interested in sciences, especially physics and maths, and they usually study for a relevant degree. Technicians, who produce calculations and drawings, train on-the-job while studying part-time. 'Technicians work for designers and, if they're good, become designers themselves,' explains Chris. A designer does some of the work of an engineer but is not as qualified. Chris started as a draughtsman and is now a chartered engineer, although almost all engineers now study for a degree rather than qualify this way. 'I'm still a technical person,' says Chris, who runs a successful engineering practice. 'Everything we do is technically based.'

Out of the office

A structural engineer would probably spend part of the week out on site. 'A good thing about this job,' says Chris, 'is its wide variety of opportunities.' Chris could be working on a £45 million building or the beam in someone's living room. 'I could be in someone's posh house working on their extension, or up to my neck in mud on a dirty site,' he laughs.

Getting started

1. Study sciences but remember the arts subjects too. Engineers need to communicate well both orally and in writing.
2. When you see a large building site, make a note of the contractor from the board outside and call them. Explain your interest and ask to be shown around the site.
3. Ask, through your school, whether you can visit the library of a university that teaches structural engineering. Look at professional magazines, research papers and textbooks.Do you understand any of it? Does it appeal to you?
4. Cruise the Internet for information about engineering and details of new building projects. See how the different professionals involved in making a building slot together.

Susan Peach – Architectural Technologist

Susan at her desk working on a final account, which is the total cost of the building works, agreed between herself and the contractor.

If you want to build an extension to your home, or a new industrial building, where do you go for advice? To a consultant in the construction industry: their roles vary but they are ready to come and work on your project. Susan Peach is an architectural technologist who works for a firm of chartered building surveyors. 'Generally we are the team leader or chief consultant, that means we are the main point of contact for the client and tell the other consultants (structural or mechanical engineers, perhaps) and the contractors (builders) what is required and when.'

A good building at the best price

Susan draws up a number of documents for her clients during a project. She starts with a feasibility study, which is a report with drawings, and the client will ok it. 'Then we progress the drawings,' says Susan, 'and they become working drawings.' To do this Susan goes back to the site and takes more details which she adds to the plans. Then she writes a specification, which details everything from the materials and workmanship to be used, right through to the health and safety precautions that must be taken on site. This is the document that goes to contractors for them to price the job. 'The more you specify, the tighter you can control cost, quality and standards,' explains Susan. 'That's what we're here for – the client gives us responsibility to supervise the works. A poor specification will produce a poor building.'

Once work starts, Susan visits the site between two and four times a week. 'We have to be able to talk to contractors, understand exactly what they are doing, and pass the information on to the client,' she explains.

Technology saves time

Susan draws on a computer using a CAD program, and then saves it onto disk before sending it to other consultants. 'Working drawings are A1 size,' explains Susan. 'But a disk only takes a couple of seconds to copy.'

An architectural technologist is similar to an architect, but concentrates more on the technical aspect of projects. 'With technical drawing you have to be spot-on, and I like that side of things. If you like technical drawing then go towards the technologist career,' says Susan. 'The work is extremely varied. You need a technical and logical mind and to enjoy working with other people. You also need to know how materials react with each other and how they might react with people: insulation can be very itchy, for instance, and can cause respiratory problems. You have to like to work outside as well as in an office and to enjoy design work. If you are designing an elevation you see how different shapes work together. Curved roofs are fashionable at the moment, but you must know which materials will curve successfully.'

Susan shows a client the progress made on her plans. Liaising with the client, discussing the costs and design of every project, is an important part of Susan's work.

Getting started

1. Use your technical drawing equipment to design a building or whatever interests you: perhaps a window detail. You might look around at some Victorian brickwork first and incorporate these designs.
2. Follow a new housing development: look in the local paper and then at notices about planning permission (often tied to lampposts); go to the planning office and ask to see the plans; you might even arrange for a small school group to visit the planning committee meeting. Then watch the building work progress. *(Note: Do not enter the building site.)*
3. Research the history of architecture. See how it has progressed through time, in urban and rural areas and in different countries.

David Wood – Micro-Brewer

We use science in every part of our daily lives, and a lot of science goes into producing the food and drink we consume. David Wood is a micro-brewer, making beer on a small scale in a brewery attached to a pub.

'Water in brewing terms is called "liquor". I add a solution of sulphuric acid to the liquor to remove the chalk.' This is only one of the many scientific reactions taking place before the mixture of water, malt and hops – plus a few extras – becomes beer. "When the local water is adjusted to the ideal brewing liquor it's called Burtonization – because Burton on Trent has ideal brewing water and I adjust my water to match theirs.'

David checks a previous brew in the cellar. It is ready to drink.

A good mash

David mixes hot liquor with cracked malt in a 'mash tun'. 'This is probably the most crucial part of the brewing process,' he explains. 'If I mess this up I would end up with a terrible beer. I'm ensuring that the malt and liquor are combined into an even consistency with the correct pH. This gets the exact balance of fermentable and non-fermentable sugars at the end of the "malt", the process of converting starch to sugars. I end up with a big load of what looks like grainy porridge.' This mash stands for 90 minutes while the naturally occurring enzymes in the malt convert starches to sugars. Then David uses more hot liquor to extract the maximum amount of sugars in a solution known as the wort. This is pumped into a vast kettle where it is boiled, with hops, to concentrate and sterilize the wort and bring out the acidity and bitterness of the hops. This acidity is not only important for flavour, but also preserves the beer.

It's not until all this has taken place, and more hops added, plus David's 'secret ingredients' for special brews, that the solution is cooled and yeast added. Then it stands in fermentation vessels for up to three days before being piped down to the cellar for conditioning.

'I have to be very practical – as well as scientifically-based. If a pump breaks down and it can't be fixed it on the spot, all the beer has to be thrown away,' explains David.

David stirs the mash.' I give this a few minutes to settle down into a decent mash. The temperature is carefully controlled; if it is too high the amount of less-fermentable sugars increases.'

Learning about brewing

After taking 'A' levels, David went straight into the pub trade. 'I've always been interested in brewing. I trained in pub management and ran a freehouse for six years. I automatically assumed that to be a brewer you needed a degree but I pestered some local brewers so much that they found me this job. A general brewer would have got a brewing degree after taking all the relevant 'A' levels,' he says. David trained under the head brewer and now runs the brewery himself.

'Brewing is a very noisy, hot, sticky and smelly job,' he laughs. If I didn't enjoy it I wouldn't do it. This is craft brewing – you can't get fresher or more unique beers than the ones produced in small plants like this. The greatest buzz for me is that people come in and spend their hard-earned cash buying something I've made. I see it right through from the recipe to the crystal pint.'

Getting started

1. Write to large breweries to find out about their training schemes.
2. Find out about other careers in the food and drink industries.
3. Talk to anyone you know in the brewing trade about recent changes in the industry.
4. Contact your local licensing magistrate or police licensing liaison officer for information about age restrictions on making, selling and consuming alcohol.
5. Write to the Brewers' Guild or the Small Independent Brewers' Association for careers and general information. Find their addresses at your library or careers service.

Classified Advertising

HELP WANTED

EXPANDING BUSINESS NEEDS PARTNERS
Full and part time.You are: ambitious, open-minded determined to SUCCEED. *No capital requirement.* Send CV to: Jan Smith, 62 Albert Street, Bigley-by-Sea, BG9 7NT

CANTONESE/ENGLISH BILINGUAL ACCOUNT EXECUTIVE
LEADING telecommunications company is currently expanding in the Oriental communities. Candidate should have at least 2 years' sales and marketin experience, insurance experi are an asset. £20 +. QUALIFIED applican Please contact Pe Manager, NCM Com East Bigley R Plumchester, PH9

SUB-EDITOR wa women's mag writing experi essential. Ple Nancy Lan Nightingale uil Church Stree H9 5

Pharm ceut

GRAPHIC TIST
Our busy A ising De and lay out s, print a
Your rela college experience t cludes fa and QuarkXP You wo priorities and work to
Please forw our CV t Trade, Summer errace, 555 4112. We a equal op

ACTIN INSTRUCT
Olivier College requires ing Instructor for the au winter academic year. This i half-time position for a ten-month contract beginning mid August.
The Instructor will teach both Introductory and Intermediate Acting for 1st and 2nd year students. Possible opportunities to direct College productions exist but is not a job requirement.
Previous teaching experience essential. Bachelor's degree, acting and directing experience required. Master's degree in a related field preferred.
Deadline for application is 25 June. Enquiries can be directed to the Personnel Department or send a CV and covering letter.
Olivier College
45 Lawrence Street
East Plumchester PH12 7FX

CITY DISPATCHER WANTED
Busy City dispatch requires innovative person to run local truck operation. Require good knowledge of ocean cargo. Apply in confidence to: GENERAL MANAGER PO Box 4329, New Amsterdam, Plumchester, PH6 2NF

SALES REPRESENTATIVE
DUTIES:
Promote a quality window products to the building market in the city.
ALIFICATIONS:
- Minimum three years' sales perience in relat
- An

will anical ap- asset. Good command English necessary.Call ABC 555 6391 for more informa
Please telephone between 9 am and 12 noon Monday to Thursday only.

SCARBIN AND BOXWELL GENERAL HOSPITAL
Registered Nurses
In-patient Services
• Full and Part-time
An active acute care facility, the new In-patient Mental Health Department, will provide the experienced mental healthcare professional or professionals motivated to pursue career in mental health nursing with an ideal opportunity to broaden their group, interviewing, and counselling skills in a supportive and cooperative setting. The background of the ideal candidate will include recent experience in primary nursing care. General qualifications include a Certificate of Competence from the Royal College of Psychiatric N

on Therap
n consultation ary nurses, you will es a vities of all types o asses patient leisure izations a lanning com- ptions are able. As the ne year of e ence in an a recogni ecreation
n and Boxw eneral onals dedicat their are invited to ly in well General H ital,

anage
ge Carpente age
e carp and ition, c idates perso elected epo ectly to
the various Philharmonic
fit programme. roduction Manager, Arts, teleph 555 7474 mit your CV and hand en letter to: Cheryl Wade, ington Centre for Perf g Arts, Vaudeville Road, mchester, PH1 6FT.

WANTED
Technical Trainee

Plumchester Optical Services require a young person to assist the technicians in this busy spectacle manufacturing unit. Duties include inputting orders, calibration, cleaning and maintenance of machinery and generally helping out. You will be expected to study relevant qualifications by day release at college, for which the company will pay all costs.
We are looking for somebody who pays attention to detail, has an interest in scientific and technical work and who can demonstrate some previous experience working with machinery or in a technical environment. Please send a CV and covering letter to:

John H Davies
Plumchester Optical Services Ltd
129 Baskerville Road
Plumchester PH7 6LT

Start Your Own Cleaning B
Be part time in cleaning shops and offices.
For more information call 555 5190

Weller
JOURNEYMAN MACHINIST
General Machinist, preferably with milling experience required for precision machine shop manufacturing electromechanical sensors. Minimum 5-10 years' related experience. Must be capable of reading detailed drawings and working to extremely close tolerances.
No telephone calls please. Forward CVs to:

Mrs P Weller
Fred Weller Corporation
34 Leslie Road
Plumchester PH8 0BT

FIRSTLINE HEALTH & FITNESS CLUB
Firstline Health & Fitness Clubs, expanding through Plumchester and the surrounding area, offers an exciting and rewarding career in the health & fitness industry. We are currently searching for sales individuals with strong interpersonal skills in the following positions.

1. Membership Consultant
2. Corporate Sales — must have corporate sales experience
3. Programme Consultant
4. Cardio Tester

Qualifications:
- Minimum 2 years' direct sales experience
- Background in aerobic & anaerobic training
- Knowledge of nutrition

Qualified applicants are invited to call Steve Florry for an information pack 555 1305 or fax your CV to 555 1304.

Who got the job?

Finding a job

The first step to success in any career is finding a job. But how do you go about finding one?

- Talk with family, friends and neighbours and let them know what jobs interest you.
- Respond to 'Help Wanted' ads in newspapers.

- Pin an advertisement of your skills on a community notice board.

- Register at Government and private employment agencies.
- Contact potential employers by phone or in person.
- Send out speculative letters and follow up with phone calls.

A job application usually consists of a letter and a curriculum vitae (CV - a summary of your experience and qualifications for the job). Applicants whose CVs show they are qualified may be invited to a job interview.

Activity

Recruiting a technical trainee

The advertisement on the opposite page is inviting applications for a junior technical post in a well-known local manufacturing company. The advertisement isn't specific about the school qualifications needed so this could be a great chance for someone who enjoys science and technical subjects but doesn't want to go on to university.

Applicants are asked to send a CV and a covering letter. Mr Davies will look at these and decide which applicants he would like to invite for interview. As there will probably be quite a few applicants it is important that applications are well thought out and neatly presented.

Two of the applicants were Trevor Manning and Liz Gonzales. Their letters and CVs and the notes made by Mr Davies during the interviews are shown on pages 46 and 47.

Procedure

Make a list of the qualities that you think are important for a good trainee in a manufacturing company. Now consider each applicant's CV, covering letter and performance during the job interview. Which candidate has the best qualifications and experience? What else, besides qualifications and experience, did you consider before making your decision?

Challenge

How would you perform at a job interview? Role-play an interview in which a friend plays the part of Mr Davies, then reverse the roles. This practice can help make sure that when you apply for a job, you have a good chance of getting it!

Trevor Manning's application and interview

4 Wolstonbury Avenue
Plumchester East
PE2 7LL

14 February 19 – -

Mr J Davies
Plumchester Optical Services
129 Baskerville Road
Plumchester PH7 6LT

Dear Mr Davies

Re: advertisement for technical trainee in Plumchester Gazette

I would like to apply for this position and enclose my curriculum vitae. As you can see, I have been interested in physics and other sciences for a long time and hope that I will have a career in this area. My holiday and temporary work at Broadloom Alarms involved sometimes helping production operatives with their technical problems as well as my official duties, and I enjoyed this very much.

I would be pleased to come in for an interview and am available any time.

I look forward to hearing from you.

Yours sincerely

Trevor Manning

Interview: Trevor Manning

* Arrived on time for interview. Wore smart, though not very clean, clothes.
* Was polite and friendly, not too nervous. Would probably fit in to the team.
* Little eye contact.
* Asked about the practical elements of the work but not about studying.
* Enjoyed his work at the alarm manufacturer and had got to know a lot of people there.
* Overall, amiable and interested, perhaps lacking motivation.

Curriculum Vitae

Name: Trevor Manning
Address: 4 Wolstonbury Avenue
Plumchester East
PE2 7LL
Telephone: 5554701
Date of birth: 11 June 19 –

Education
19 – - – 19 – Plumchester Secondary School
7 GCSEs grade A-C including technical drawing, physics, and chemistry.

Work experience
19 – Summer work with neighbour, gardening, mending fences, etc
19 – Saturday and summer work at Broadloom Alarms
19 -- Six month temporary contract with Broadloom Alarms

Interests
Mechanical: I have always enjoyed machines of all sorts and I help out at a local steam railway, doing odd-jobs in the engine sheds.
Social: I have a wide range of social activities and attend the Plumchester Lions rugby club, where I am in the second team.
Reading: I enjoy reading about the history of steam and other engine-related topics, and listening to music.

References available on request

Liz Gonzales's application and interview

14 February 19 –

32 The Ridgeway
South Plumchester
PH12 1RR

Mr J Davies
Plumchester Optical Services
129 Baskerville Road
Plumchester PH7 6LT

Dear Mr Davies

Re: technical trainee

I saw your advertisement for a technical trainee and I am very interested in the post. I have always enjoyed technical things and have shown a flair for them at home and school. I have been considering whether to go to college full-time but would much rather train on-the-job and gain qualifications at the same time. I am a keen worker and willing to start at the bottom and learn.

I enclose my CV and would be very pleased to tell you anything else you might need to know. Just call me at home if necessary.

I am looking forward to hearing from you.

Yours sincerely

Liz Gonzales

Curriculum Vitae

Elizabeth Gonzales
32 The Ridgeway
South Plumchester
PH12 1RR
Telephone 5554722

Education
Plumchester Secondary School 19 – – 19 –
8 GCSEs at grades A–C
Lower sixth, first year of Advanced GNVQ
Favourite subjects: all sciences, technology, art, and music

Work experience
I have worked in a yacht chandlers and a DIY superstore, advising customers on the best purchases for their needs. I am the DIY person at home and mend all sorts of items. I have a weekend job at a light fittings factory, doing production work, and this involves sometimes helping out in the warehouse. I gain a wider picture of the whole of the business there.

Interests
All engineering projects; car maintenance; reading; sport; socializing; listening to and playing music.

References: available on request

Interview: Liz Gonzales

- Arrived early and chatted to the receptionist.
- Wore appropriate smart clean clothes and her hair was well tied back.
- Knew a little about the business and asked a lot of questions.
- Is keen to learn about optics.
- Is not as knowledgeable about technical issues as she thinks.
- Very personable.
- Keen to study.
- Overall, a good team member although unsure about technical expertise.

Index

Credits

(l=left; r = right; t = top;
b = bottom; c = centre;
bl = bottom left;
br = bottom right)

All photographs by Joanna Grigg, except Page 8: (t) The Automobile Association
Page 8: (c) The Automobile Association
Page 14: University College London (UCL)
Page 20: BT Corporate Picture Library; a BT photograph
Page 21: Royal College of Speech and Language Therapists
Page 23: (t) BT Corporate Picture Library: a BT photograph
Page 23: (b) BT Corporate Picture Library: a BT photograph
Page 25: (b) BT Corporate Picture Library: a BT photograph
Page 26: BT Corporate Picture Library: a BT photograph
Page 31: The Association of the British Pharmaceutical Industry (ABPI)
Page 37: Phil Smith
Page 40: Vivienne Parry

Answers to quiz on page 13

Here are examples of the light sources named

1. Light emitting diodes – the red light indicating that a tape recorder is on.
2. Incandescent light – light bulbs where a wire heats up.
3. Fluorescent light – fluorescent tubes, often used in kitchens, schools and offices.
4. Light created when electrons hit a phospher – a television.
5. Lasers – in CD players (trick question: the light is infra red so we can't see it, though newer technology means we soon will.)
6. Sodium bulbs – street lights.
7. Nuclear power – the sun and stars.
8. Electroluminescence – sometimes, although rarely, seen lighting the back of LCDs.